HIDEKI YUKAWA
"TABIBITO"
(The Traveler)

Translated

by

L. Brown & R. Yoshida

World Scientific

ISBN 9971-950-09- X
ISBN 9971-950-10-3 pbk
Cover designed by Galen Song
Typesetted by AZ Graphics
Printed by Singapore National Printers (Pte) Ltd.

Contents

Foreword

My fiftieth birthday occurred in January last year;[1] on that day I had lived just half a century.

The path I had taken was not a difficult one, measured by the usual standards. I was born in a scientist's family and grew up with brothers who each became a scientist in one or another area; my education was broadly liberal; I did not toil in worldly ways. I had a fortunate environment.

But my academic life is not so easy to analyze. While I was lucky in some ways, it cannot be denied that I experienced greater hardships than some others. Physics is a science that has made rapid progress in the twentieth century. It could be said that I simply rode with the incoming tide of a new science doing what I liked to do as I liked to do it. Nothing certain can be said except that I desire, as I did in the past, to be a traveler in a strange land and a colonist of a new country.

Sometimes a colony that once yielded a rich harvest is cast aside. Today's truths may tomorrow be disproved, and that is

[1] January 23, 1957

why, from time to time, we must look backwards in order to find the path that we must take tomorrow.

I have spoken of two paths, but in actuality there is but one. For the path I took as a scientist is the same one I took as a person. During the past twenty years, I have written about my past in several brief essays. Others have written various things about me; there are several biographies. The public already has a certain image of me as a person, and I want to offer some material that can be used in judging that image.

When a person looks in the mirror, he sees the face that is seen by others. Yet a listener may be surprised when he reveals the inner self that is not seen by others. The two different views may be hard to reconcile; especially so in my case, because I have always had trouble expressing myself. Also, I tend to view matters subjectively, and if I try to be objective I may betray myself.

In any case, not even I can perceive clearly what is about to take shape. The publisher (Asahi Shinbun) has given me a chance to fulfill the desire that came to me near my fiftieth birthday. During the past year, I have worked on this book in my spare time; two months ago I had my fifty-first birthday.

I plan to write about my relatives, within limits; my friends and teachers will also take part. A large part of this memoir should be called *Hideki Ogawa and his surroundings,* rather than *The autobiography of Hideki Yukawa,* for Ogawa is the name of my father's house.[2]

Hideki Ogawa was born in 1907 at old Tokyo's Ichibei-cho Azabu. The house smelled of plum blossoms each spring.

---

[2]Hideki Ogawa assumed his wife's family name upon his marriage.

Chronological Table

1907 Jan. 23	Hideki Ogawa was born in Tokyo, Japan.
1908 (1 year old)	His father Takuji Ogawa became professor of geography at Kyoto Imperial University and he moved to Kyoto with his family.
1913 (6)–1919 (12)	He went to Kyogoku Primary School, Kyoto.
1919 (12)–1923 (16)	He went to Kyoto Prefectural First Middle School.
1923 (16)–1926 (19)	He was a student of The Third High School.
1926 (19)–1929 (22)	He studied physics in the Faculty of Science, Kyoto Imperial University.
1929 (22)–1932 (25)	Unpaid junior research associate of Kyoto Imperial University.

1932 (25)   He married Sumi Yukawa and became a member of the Yukawa family.

1932 (25)–1934 (27)   Lecturer of Kyoto Imperial University.

1933 (26)–1936 (29)   Lecturer of Osaka Imperial University.

1934 (27)   He delivered a talk "On the Interaction of Elementary Particles I" at the Regular Monthly Meeting of the Physico-Mathematical Society of Japan held at the Imperial University of Tokyo on November 17, in which he proposed a new field theory of nuclear forces and predicted the existence of the meson. He submitted an article on this theory for publication to the "Proceedings of the Physico-Mathematical Society of Japan."

1935 (28)   The above article was published in the Proceedings.

1936 (29)–1939 (32)   Associate Professor of Osaka Imperial University.

1936 (29)   His first book "Theory of Beta Decay" was published in Japanese.

1937 (30)   He published "On the Theory of Elementary Particles II" with Shoichi Sakata.

1938 (31)   He published "On the Theory of Elementary Particles III and IV" in collaboration with Shoichi Sakata, Mituo Taketani and Minoru Kobayashi. He received a Ph.D. from Osaka Imperial University.

1939 (32)–1970 (63) Professor of Kyoto Imperial University. (The university was renamed Kyoto University in 1947).

1939 (32) He visited Europe, on invitation, to "The Solvay Conference on Elementary Particles and their Interactions" but, due to the outbreak of World War II, the conference was cancelled. Then he traveled to U.S.A. where he met many physicists and delivered lectures on the meson theory.

1940 (33) He was awarded the Imperial Prize of the Japan Academy.

1942 (35) He published a series of articles "On the Foundation of the Theory of Fields" in Kagaku (Science) in Japanese. These articles were the first publications of his efforts to construct the quantum field theory without divergent difficulty.

1943 (36) He was awarded the Order of Decoration of Cultural Merit of Japan.

1946 (39) He founded an academic journal in European languages: "Progress of Theoretical Physics".

1948 (41)–1949 (42) Visiting Professor of the Institute for Advanced Study in Princeton, USA.

1949 (42)–1953 (46) Professor of Columbia University in New York, USA.

1949 (42)   He received the 1949 Nobel prize in physics for the meson theory of nuclear forces.

1953 (46)   He returned to Japan.

1953 (46)–1970 (63) Director of Research Institute for Fundamental Physics, Kyoto University.

1955 (48)   He was one of the 11 signatories to the Russell-Einstein Manifesto against the use of nuclear weapons and for the abolition of war.

1956 (49)–1957 (50) A member of Atomic Energy Commission of the Japanese government.

1957 (50)   He attended "The First Pugwash Conference on Sciences and World Affairs" held at Pugwash, Canada.

1961 (54)–1965 (58) President of the World Association of World Federalists.

1962 (55)   He organized "The First Kyoto Conference of Scientists" for nuclear disarmament with Sin-itiro Tomonaga and Shoichi Sakata.

1965 (58)–1981 (74) Honorary President of the World Association of World Federalists.

1968 (61)   He published "Field Theory of Elementary Domains and Particles, I and II" in collaboration with Yasuhisa Katayama and Isao Umemura. These were his last scientific papers.

1970 (63)   He was granted the title of Professor Emeritus of Kyoto University when he retired from the university.

1973 (66)   He published a collection of his principal essays "Creativity and Intuition" in English (Kodansha International Ltd., Tokyo, New York and San Francisco).

1979 (72)   He published his collected papers "Hideki Yukawa Scientific Works" (Iwanami Shoten, Publishers, Tokyo).

1981 (74)   He organized "The Fourth Kyoto Conference of Scientists".

1981 Sept. 8   He passed away.

(written by Michiji Konuma)

Yukawa received the 1949 Nobel prize in physics from the hands of the Crown Prince of Sweden, December 10, 1949.

As the director of the Research Institute for Fundamental Physics, Kyoto University, around 1960.

Introduction: Hideki Yukawa and the Meson*

When Hideki Yukawa graduated from Kyoto University in 1929 and began research on nuclear theory, the knowledge of nuclear physics accumulated since the start of the century had led to a crisis. It was believed that electrons and protons must be the constituents of the nucleus, for they were observed, in special circumstances, leaving the nucleus, and because they were the only known elementary material particles — yet this picture contradicted quantum mechanics.

The solution was thought to lie in a new dynamics, or in a modified structure of space and time. Instead, major problems were resolved by the discovery of two new elementary particles, the neutron and the positron, and by Pauli's invention of the hard-to-observe neutrino. Based upon these new particles, Werner Heisenberg and Enrico Fermi made theories which provided a phenomenological basis for the strong nuclear binding force and for nuclear β decay. A deeper foundation for these theories was given by Yukawa in the form of the meson theory.

* Adapted from a lecture by L.M. Brown at the European Centre for Nuclear Research (CERN), Geneva, Switzerland, Oct. 1, 1979. For a more complete account, see by the same author, "Yukawa's Prediction of the Meson," *Centaurus 25* (1981), pp. 71–132.

The meson played a dual role: it carried energy, momentum, and electric charge between neutron and proton, producing a strong nuclear force; it also decayed weakly (i.e., with small probability) into electron and antineutrino (if electrically negative), or positron and neutrino (if electrically positive) — Yukawa's mechanism for nuclear beta decay. While mesons could never be emitted in ordinary nuclear transformations, they could materialize in processes involving sufficient energy release, such as the interactions of cosmic rays. Produced this way, the mesons would either be absorbed by matter, or would themselves decay by the beta process.

The theory thus predicted the existence of a new type of elementary particle that could be produced in a free state only in a regime of much higher energy than was available in the nuclear laboratory at that time. Yukawa's work thus went beyond the theory of nuclear forces and directed attention toward the field of high energy, or elementary particle, physics.

As described in the pages that follow this introduction, Yukawa was born in Tokyo on January 23, 1907 as Hideki Ogawa.[a,b] A year later he moved with his family to Kyoto when his father, Takuji Ogawa, left his staff position in Tokyo with the Bureau of Geological Survey, to become Professor of Geography at Kyoto University. Yukawa's father was of the same generation as Hantaro Nagaoka, Japan's most famous physicist of the pre-quantum era. Nagaoka's early work on atomic models attracted the attention of the Cambridge school and Ernest Rutherford, and he became an authority on spectroscopy, geophysics, and other subjects. On one occasion he said

a. The family name Yukawa was assumed by Hideki in 1932 when he married Sumi Yukawa and was adopted by her father Genyo, an Osaka physician. The practice of adopting the younger son of one family into a family without a son is quite common in Japan; Hideki Yukawa's father, Takuji had also been adopted in this way.

b. The period described by this book is from childhood to 1935, the year of publication of Yukawa's first article on the meson. Other autobiographical material by Yukawa in Japanese: *Butsurigaku ni kokorozashite (Aspiring for Physics; Kyoto, 1944); Meni mienai mono (of things that cannot be seen; Kobunsha, 1942); Shinri no ba ni tachite (In the course of our study),* with S. Sakata and M. Taketani (Mainichi Shimbun, Tokyo, 1951).

that "there was no point ... to be born a man, if I failed to enter the advanced ranks of researchers and to contribute to the development of some field of learning." Late in life Yukawa said that "probably the one decisive factor" to set him on the path of physics research, when he was still in high school, was that "one could find among the Japanese ahead of one such a great physicist".[1] When Yukawa got his first job, as Lecturer at Osaka University, Nagaoka was its president.

Before he started school Hideki began to study the Chinese classics with his grandfather by a method of instruction, called *sodoku*; it consisted of reading the Chinese characters (*kanji*) aloud in Japanese pronunciation without attending to the meaning of the text. Thus he learned to read very early, and he read very widely as a child. He especially enjoyed imaginative literature, including Japanese and Western classics. Among Japanese authors he mentions Natsume Soseki, whose 1908 novel, *Sanshiro*, concerns a lad from the provinces who attends Toyko Imperial University. At a gathering of students, one of them makes a public speech along these lines:

"We, the youth, can no longer endure the oppression of the old Japan. Simultaneously, we live in circumstances that compel us to announce to the world that we, the youth, can no longer endure the new oppression from the West. In society, and in literature as well, the new oppression from the West is just as painful to us, the young men of the new age, as is the oppression of the old Japan."[2]

In adolescent search for life's meaning, Yukawa discovered the writings of the Taoist sages Laotze, Chuangtze, and later Motze. These authors, unlike Confucius, placed nature, not man, at the center of the universe. He said later; "Theirs is a type of fatalistic naturalism very much like that to which the scientific view of nature may ultimately lead."[3] Yukawa, partly in rebellion, was thus attracted to a style of thinking that was materialist and dialectical, but that was different from the dialectical materialism of Marx and Engels that strongly influenced his first students and collaborators. When asked about Yukawa's philosophy recently, one of them, Mituo Taketani, responded, "Well Yukawa is a genius-type, and so he has his own philosophy." Although interested in hearing about Marxist philosophy and Taketani's *own* methodology of science, Yukawa says little about these subjects in his essays, public talks, and published dialogues.

In 1923 Yukawa entered the Third High School of Kyoto, an institution more like a German *Gymnasium* or French *lycée* than an American high school. In 1926, he took and passed the entrance examination of the Department of Physics of Kyoto University and began his studies there. One of his high school classmates, who also went to Kyoto University to study physics, was Sin-itiro Tomonaga, the son of a philosophy professor at the University. Tomonaga and Yukawa helped each other learn quantum mechanics; after graduation they stayed on together for several years as unpaid assistants at the Physics Department. In 1932 Tomonaga moved to Tokyo to join the group of Yoshio Nishina at the Institute for Physical and Chemical Research, a private research foundation, while Yukawa remained at Kyoto University as Lecturer of Physics. Tomonaga (who died in 1979) shared the Nobel Prize for Physics in 1965 with Julian Schwinger and Richard P. Feyman "for their fundamental work in quantum electrodynamics." In *Tabibito* Yukawa contrasts himself as a stubborn person tending "to go too far" without enough thinking, with Tomonaga, who was more controlled — "a person aware of the limits, who yet comes up with clever ideas."

In 1974 Yukawa gave a lecture in which he discussed the natural philosophy that prevailed when he graduated from Kyoto University:

> At this period the atomic nucleus was inconsistency itself, quite inexplicable. And why? — because our concept of elementary particle was too narrow.

> There was no such word in Japanese and we used the English word — it meant proton and electron. From somewhere had come a divine message forbidding us to think about any other particle. To think outside of these limits (except for the photon) was to be arrogant, not to fear the wrath of the gods. It was because the concept that matter continues forever had been a tradition since the times of Democritus and Epicurus. To think about creation of particles other than photons was suspect, and there was a strong inhibition of such thoughts that was almost unconscious.[4]

In the same lecture he stressed the inadequacy of the proton-electron nuclear model, evidenced by the notorious violation of the spin and statistics theorem of quantum mechanics. For example, in the proton-electron model, the nitrogen-14 nucleus would contain 14 protons and 7 electrons, or 21 fermions; yet the molecular spectrum of nitrogen gas showed that it behaved as though it contained an *even* number of fermions.

Aside from the problem of understanding the nucleus, the other major challenge to theoretical physics, as seen by young Yukawa, was to make a theory of photons interacting with electrons in a relativistically self-consistent way. Yukawa liked to refer to the problem of relativistic quantum field theory as a "settling of accounts". By this he meant that after reaping the great rewards of quantum theory in treating non-relativistic mechanical systems (atoms, molecules, and crystals), theoretical physics was then morally obliged to try to solve that old puzzle of the quantum theory: the wave-particle duality of the photon.

The leading theorists of the day were prepared to accept quite radical hypotheses in order to solve the nuclear problem. Indeed, it seemed that at the nuclear scale of distance (the same as that at which H. A. Lorentz had, early in the century, predicted a breakdown in electromagnetic theory), the triumphant revolution of quantum mechanics was about to be repeated, and that a new fundamental generalization of dynamics might be realized. Bohr attributed the peculiarities of beta decay to a failure of energy conservation; Heisenberg sought to introduce a new fundamental constant, a quantum of length to characterize the nuclear region.[5]

In the spring of 1931, Yoshio Nishina (1890–1951) of the Institute of Physical and Chemical Research (IPCR) of Tokyo,[c] lectured on quantum mechanics at the University of Kyoto. The lectures were intended for the professional staff rather than students, and consisted of an introduction to Heisenberg's *Die physikalischen Prinzipien der Quantentheorie (1930)*. Nishina is properly regarded as the father of nuclear and cosmic ray physics in Japan. A graduate in Electrical Engineering of Tokyo University, he joined IPCR one year after it was founded in 1917, with H.I.H. Prince Fushimi at the head of its Board of Trustees. After several years he was sent abroad for further study and research: one year at the Cavendish Laboratory in Cambridge, England, one year at Göttingen, and six years at Copenhagen with Niels Bohr, where he wrote a famous theoretical paper with Oskar Klein on the rate and angular distribution of Compton scattering. Returning to Japan in 1928, he began to build the Nishina Group at IPCR in Tokyo, primarily to do research in nuclear physics.

The lectures at Kyoto, and perhaps to a greater extent the social contact with Nishina, transformed Yukawa. Shoichi Sakata, who was a relative of Nishina by marriage, often visited him and was introduced on one of these occasions to both Yukawa and Tomonaga, who were having intense discussions with Nishina on nuclear physics. Yukawa said later that Nishina "carried with him the spirit of Copenhagen." Furthermore, Nishina's personality put Yukawa at ease; usually silent and withdrawn, he found that he could talk with the older man.

The year 1932 was a turbulent one for Hideki Ogawa. In that year he married Sumi Yukawa, took her family name, and went

---

[c]The Japanese name is *Rikagaku Kenkyusho*, or *Riken* for short.

to live in her family's home in crowded Osaka, a busy commercial seaport very different from his own restrained and traditional Kyoto. He was appointed Instructor in the University of Kyoto and asked to lecture on quantum mechanics. To the shy young man, who still wore the short-cropped hair and the school uniform of a student, so many changes were upsetting. Among the students attending his first lectures were Sakata and Kobayashi, followed the next year by Taketani. These three later became his collaborators in developing the meson theory after the discovery of the cosmic ray meson in 1937.

As Yukawa points out in his autobiography, 1932 was even more turbulent for nuclear physics than it was for his personal life, being marked by the discoveries of the deuteron, the neutron, and the positron, and the disintegration of nuclei by artificially accelerated protons. Yukawa became aware of the opportunity to make a truly significant contribution to theoretical physics. Before the discovery of the neutron permitted a nuclear model without electrons, Yukawa had had the same difficulties with it as other physicists. He studied George Gamow's *Constitution of Atomic Nuclei and Radioactivity* (1931), where many discussions dealt with the properties of "nuclear electrons". These passages were set off between special signs (intended to be skull and crossbones originally) to remind the reader that the content was dangerously speculative.

One such passage deals with the nuclear beta decay, where electrons appear to emerge directly from the nucleus, apparently proving that the nucleus contained electrons. However, Gamow says,

> These results lead us to a very strange conclusion. Since there is no process compensating for the difference of energy lost by different nuclei of the same element in the ejection of a β-particle, we must deduce, according to the principle of conservation of energy, that the internal energy of a given nucleus can

take any value within a certain continuous range. This
... however, has not the slightest effect before or after
the β-emission ... there is no trace of a continuous
distribution of energy in the emission of α-particles or
γ-rays. In these processes all the nuclei seem to be
again identical.

Perhaps, Gamow concludes, "as was pointed out by N. Bohr,
we must reckon with the possibility that the continuous distribution of energy among the nuclei is fundamentally not obser-
vable, or, in other words, has no meaning in the description of
the physical processes ... *This would mean that the idea of
energy and its conservation fails in dealing with processes involving the emission or capture of nuclear electrons.*"[6]

Although Pauli had begun to suggest the possible existence of
a neutrino (he called it "neutron") by the end of 1930, he did so
very cautiously and without publishing it.[7] Yukawa said that he
was unaware of Pauli's thinking on beta decay, but knew
Bohr's ideas from reports of the 1931 International Conference
on Nuclear Physics, held in Rome.

It was in 1932 that Heisenberg published an important work
(in three parts) proposing a neutron-proton model of the atomic
nucleus. A certain confusion still surrounds Heisenberg's
papers, arising from his attempt to combine two immiscible
theories. One was an essentially correct theory of *nuclear*
structure; the other was an essentially incorrect theory of
neutron structure. The former, in its purely phenomenological
aspect, treated the neutron as an elementary particle of spin
one-half, having small magnetic moment, and obeying Fermi-
Dirac statistics; the latter, a fundamental theory of the neutron,
treated it incorrectly as a proton-electron compound. This
structure of the neutron seemed necessary to account for the
existence of the neutron and to allow for beta decay. Other
observations also seemed to require a composite neutron: one
was the "anomalous absorption" of hard gamma rays; the
other had to do with cosmic-ray electrons. These processes were

mentioned by Heisenberg at the end of Part I of his nuclear structure paper (where he said that his theory *does not hold* for them), and they were further discussed in Parts II and III. He argued that even alpha particles must be made of protons and electrons, in order to have the large polarizability required to explain the gamma ray phenomena.[8] In fact, the high energy processes mentioned turned out to be *electromagnetic* in nature, and in no way related to the continuous β-ray spectrum.

Heisenberg discussed the consequences of the assumption that atomic nuclei are built out of protons and neutrons, without "loose" electrons (those not bound in neutrons). This reduced the β-decay problem to the question of how the *neutron* in the nucleus decays into proton and electron. "To interpret the neutron as composed of a proton and electron, one would have to ascribe to the electron Bose statistics and spin zero. However, it does not seem useful (zweckmässig) to consider this picture more closely." Rather, "the neutron should be considered an independent fundamental particle that can, under appropriate circumstances, split into a proton and electron, whereby presumably the conservation laws of energy and momentum are no longer applicable." The violation of these conservation laws was not a consequence of Heisenberg's nuclear model, but was assumed in order to fit the β-decay observations without using the neutrino, while angular momentum conservation and the spin-statistics theorem *necessarily* fail in Heisenberg's model, even in the nuclear binding problem.

Heisenberg introduced analogies drawn from molecular physics that resulted in his famous charge-exchange force. The strongest interaction, by analogy with the H_2^+-ion, was supposed to be that between proton and neutron. The electron was associated in a quasi-atomic state (neutron) with one or another proton alternately. In Heisenberg's model, the neutron was thus viewed as a closely bound state of a *spinless* electron obeying *Bose* statistics. However, he used the exchange force as phenomenological, "without reducing it to electronic motions."

The composite picture of the neutron led also to a neutron-neutron force (analogous to the force in the neutral hydrogen molecule where two protons share a pair of electrons) which was weaker than that between proton and neutron. Both exchange forces were expected to vanish for separations larger than about 10^{-12} cm. Since Heisenberg considered the proton as a fundamental particle, only the basic Coulomb repulsion acted between protons. Finally, in order to complete the energy balance, a term was added in the Hamiltonian to account for the mass difference between neutron and proton.

Applying his theory, Heisenberg showed that the helium nucleus should have "closed shells" of both protons and neutrons, and should be particularly stable. Structures containing only neutrons would not be bound so strongly as structures having at least some protons. Thus, the beta-decay process would occur in nuclei that were too "neutron-rich". As the positron had not been discovered at the time of Part I, he did not discuss the possibility of positron decay of "proton-rich" isotopes. Indeed, in Heisenberg's theory, where the proton was *not* composite, no mechanism existed for it.

On the basis of strong forces of short range (and with modifications by Eugene Wigner and Ettore Majorana) a reasonable picture of the nucleus emerged, yielding at least a qualitative understanding of the nuclear energy level spectrum and of elastic and inelastic scattering of nuclear projectiles by the nucleus. With regard to radioactivity, no important conceptual problems remained concerning alpha and gamma radioactivity, but without the neutrino it remained impossible to construct a consistent theory of beta decay. So long as beta decay was regarded as a process in which an electron emerged from the nucleus with a *variable* amount of energy, while the nucleus made a transition between *definite* quantum states, it was impossible to maintain the conservation of energy and momentum.

Nevertheless Heisenberg, Bohr, and others, *rejected* the

neutrino, failed even to mention it! Before admitting the possible existence of a new elementary particle, they were willing to abandon the hard-won conservation laws that most physicists consider the very pillars of physical science. Bohr and Heisenberg, indeed, seemed to welcome the apparent failure of quantum mechanics at nuclear dimensions, and hoped that it might lead to radically new conceptions of space, time and causality.

But while some who had made the modern quantum theory were plotting a second revolution, Enrico Fermi made a "tentative theory of beta rays" that conserved all the important dynamical quantities through the simultaneous creation and emission of a neutrino together with the beta decay electron.[9] His theory complemented Heisenberg's nuclear model; by using both, the nucleus could be regarded as an entirely quantum mechanical system. Fermi assumed the existence of Pauli's neutrino, and he allowed only protons and neutrons to be in the nucleus. Consequently electrons and neutrinos had to be created when emitted; in that respect, they resembled the photons emitted by an excited atom or nucleus.

To deal with the variable number of light particles, it is necessary to use the method of quantum field theory, which is often called "second quantization". This is analogous to the quantization of the classical Maxwell electromagnetic field that represents it as a collection of identical elementary particles, the photons. However, in the case of electrons, the "field" that is to be quantized is not really a classical field, but is the Dirac wave function "field" that represents the *first* quantization. (Hence the name "second quantization".) For the β-decay electrons and neutrinos, both the wave function theory and its second quantized version must be relativistic and must incorporate the Pauli exclusion principle for the Dirac particles — not the symmetric (Bose-Einstein) statistics that light quanta obey. Fermi recognized that various combinations of the field components of electron and neutrino would allow the interaction Hamiltonian to be Lorentz-invariant, so his theory was not uniquely determined. Subject to possible later modification, he chose those particular combinations which transform as a Lorentz four-vector, as does the electromagnetic four-potential. In the heavy particle Hamiltonian, where normally the electromagnetic four-potential would appear, Fermi placed the electron-neutrino four-vector field and obtained a familiar looking Lorentz-covariant interaction.

Since beta decay half-lives are very long on the atomic scale, Fermi knew that the coupling constant, analogous to the electric charge, was small enough to justify the use of perturbation theory, and calculated the half-lives and the decay-electron spectra for a number of radioactive elements. He showed that the shape of the electron spectrum at its high energy endpoint depends upon the mass of the neutrino, and inferred that the value of that mass was probably zero.

After the neutron discovery, when Heisenberg's papers on nuclear structure arrived in Japan, Yukawa's interest in nuclear physics quickened. Sakata, then a third year student at Kyoto, recalled that Yukawa advised him to write his bachelor's thesis on that subject. Yukawa translated Heisenberg's papers and wrote a perceptive introductory essay, his first publication, in which he recognized that Heisenberg's theory was really phenomenology and needed a better fundamental mechanism. Soon afterwards, Yukawa presented a paper, entitled "Theory of Nuclear Electrons," at a meeting of the Physico-Mathematical Society of Japan, held at Sendai in April of 1933.

During the Sendai meeting, Yukawa's brother introduced him to Hidetsugu Yagi, who had just become Head of the Physics Department in the new Science Faculty of Osaka University. He was offered a Lectureship, and during the next year Yukawa divided his time between Kyoto and Osaka Universities. In April, 1934 he moved into the new Physics Building in downtown Osaka as full-time Lecturer and head of the "Yukawa Group". In the same month Seishi Kikuchi came from Tokyo IPCR to do experimental nuclear physics and

Sakata, who had been working in Tokyo with Tomonaga, came and shared an office with Yukawa.

It may have been Heisenberg who first suggested a connection between exchange forces and what became known as the "Fermi field", an analogue of the electromagnetic field, with the electron-neutrino pair substituting for the photon. In letters to *Nature* shortly after the appearance of Fermi's β-decay theory, the Russian physicists Tamm and Iwanenko estimated the effective potential arising from the exchange of a pair of light particles.[10] They found that it falls off with distance as the inverse fifth power when the distance is small. However, the resulting force is too weak to provide the neutron-proton binding force by a factor of at least 10^{10}.

Yukawa said that he first came upon Fermi's β-decay paper in 1934, and grasped immediately that the electron-neutrino pair could provide a nuclear force that would be free of the contradictory aspect of mere electron exchange, and that would preserve the conservation laws in nuclear beta decay. When Tamm and Iwanenko showed that the force of electron and neutrino exchange was far too weak to account for the nuclear force, Yukawa recalled another idea that had earlier entered his consciousness, and he said,

> "Let me not look for the quantum of the nuclear force field among the known particles — including the neutrino. If I pursue the characteristics of the nuclear force field, then the nature of the quantum of that field must also become apparent."

One night in October 1934, shortly after his wife had given birth to their second child, unable to sleep, Yukawa suddenly saw that the range of the nuclear force must be inversely proportional to the mass of its field quantum, and that there must be a new particle. In the morning he found the mass to be about 200 electron masses, and concluded that it should appear both positively and negatively charged. He reported this at his usual lunch with the Kikuchi group. Kikuchi remarked that the particle should be visible in the Wilson cloud chamber; Yukawa agreed — and argued that it should be found in cosmic rays. Not long afterwards the theory was presented at the Osaka branch meeting of the Physico-Mathematical Society of Japan, and a month later at the Tokyo Imperial University.

Yukawa had been thinking for several years about the problem of the nuclear force field; it appeared to him later that he had glimpsed the solution repeatedly without arriving at a definite formulation of it. When he did grasp the essential relation between the quantum mass and force range and sat down to write his article, it took barely a month to complete the work. It contains several new and valuable ideas, and its physical logic proceeds inexorably.[11]

In his "meson" article, Yukawa argues that the interactions of Fermi and Heisenberg are different, that they give alternative ways for a neutron to become a proton (or the reverse): either an electron-neutrino pair is emitted (Fermi process), or the energy and negative electric charge are taken up directly by another proton in the nucleus, which thereby becomes a neutron (Heisenberg process). The first occurs with small probability (i.e., it is a weak interaction); the second has a probability large enough to provide the binding of nucleons in the nucleus (i.e., it is a strong interaction).

Yukawa thus takes a modest step backward, disassociating β-decay from nuclear binding forces, but now focuses his attention upon a new *field of force*, which like gravitation and electromagnetism — is universal, interacting strongly with heavy particles and weakly with light particles. His final step in the reasoning that leads to the meson theory is:

> In the quantum theory this field should be accompanied by a new sort of quantum just as the electromagnetic field is accompanied by the photon.

The parallel with electromagnetism is exploited by expressing the force between neutron and proton in terms of a potential: "The potential of force between the neutron and proton should, however, not be of Coulomb type, but decreases more rapidly with distance. It can be expressed, for example by,

$$+ \text{ or } -g^2 \exp(-\lambda r)/r$$

where g is a constant with the dimension of electric charge ...". The constant λ has the dimensions of an inverse length and determines how fast the potential falls off with the distance r. The reciprocal of λ is called the *range* of the potential, for the potential falls off very rapidly at distances greater than $1/\lambda$. One adjusts λ to obtain the nuclear force range, about 10^{-13} cm.

Yukawa pursues the electromagnetic parallel: the potential can, in general, be time-dependent. Furthermore, recall that the electric and magnetic fields are derived not from a scalar potential alone; there is a vector potential as well, which together with the scalar potential forms a relativistic four-vector. The *source* of the electromagnetic four-potential is the charge-current density four-vector; Yukawa needs to write the analogue of this source. Finally, the theory must be quantized; then one gets the quanta of the nuclear field, the analogue of photons, the mesons.[d] Since the emission of a meson involves a change of neutron to proton, or vice versa, the nuclear field quanta must carry electrical charge, unlike the photons.

Yukawa's energy function for a pair of nucleons interacting by meson exchange resembled Heisenberg's, with this difference: where Heisenberg had an unspecified function of the separation that he called "exchange-integral", but actually regarded as something to be determined empirically, Yukawa placed $-(g^2/r) \exp(-\lambda r)$, the now-famous Yukawa potential. At first Heisenberg chose the sign of the exchange integral to be that of his molecular analogy, but this turned out to give the

---

[d]Yukawa called his particles "heavy quanta" or "U-quanta". The name *meson* refers to the particle's mass, intermediate between that of the electron and the nucleon.

wrong spin for the deuteron. For this reason, Yukawa gave his potential the *opposite* sign; but here he made an error. A fundamental field theory does not permit a sign to be freely chosen — it predicts it. Yukawa's deuteron would also have the wrong spin, and later it became evident that other versions of the meson theory (e.g., other spins) would need to be investigated.

The theory to this point is semi-classical; i.e., it is a quantum mechanical description of nucleons interacting through a *classical* U-field. However, there is another interpretation for the wave equation for U: it can be regarded as the relativistic Schrödinger equation for a free particle, so that U is the particle's *wave function*. The range parameter λ is then the inverse Compton wave length of the particle, mc/h, and gives the mass of the quantum.

Assuming a nuclear force range of 2×10^{-13} cm, Yukawa predicts that the mass of the U-quantum (of either charge) should be about 200 electron masses. He notes,

> As such a quantum with large mass and positive or negative charge has never been found by the experiment, the above theory seems to be on a wrong line. We can show, however, that, in the ordinary nuclear transformation, such a quantum can not be emitted into outer space.

However, where sufficient energy can be transferred in an elementary process, e.g., in the cosmic ray interactions, U-quanta *should be observed*.

These are the possibilities that arise from the strong inter-action of the U-quantum with the nucleons:

(a) A neutron becomes a proton, emitting a negatively charged quantum; this quantum is absorbed by a proton in the same nucleus, the proton becoming a neutron.

(b) A proton becomes a neutron, emitting a positively charged quantum; this quantum is absorbed by a neutron in the same nucleus, the neutron becoming a proton.

(c) The same processes described by (a) and (b), but with the second interaction occurring in another nucleus. This provides a force acting between nuclei, responsible for nuclear scattering and nuclear reactions.

(d) The same production processes as in (a) and (b), with the U-quantum emitted and moving as a free particle.

In case (d), one can ask for the ultimate fate of the U-quantum. Evidently, it may have a strong reaction at a nucleus far removed from its birthplace, or it may undergo some other transformation as a free particle in free space. (Figures 1, 2, 3.)

Yukawa assumes that the U-quantum (or meson) interacts not only with the "heavy particles", proton and neutron, but also with the "light particles", electron and neutrino (actually antineutrino) and their antiparticles. As with the nucleons, the light particles interact as a pair, having one charged and one neutral member. A negative meson can be absorbed by the "vacuum", and raise a negative energy neutrino to a positive energy *electron* state, leaving a hole in the Dirac "negative energy sea" of neutrinos — this hole is an *antineutrino*. The process occurs with a much smaller probability than the interaction of the meson with nucleons; it can be regarded as the radioactive β-decay of the U-quantum, when the latter is a free particle (Fig. 3). On the other hand, if the process takes place in the same nucleus in which U-quantum is produced, it is regarded as the β-decay of the nucleus (Fig. 2).

Because of the short range of the interactions, Yukawa's β-decay theory "does not differ essentially from Fermi's", as Yukawa states explicitly. On the other hand, it has a most important new consequence. Because the meson plays an essential role in β-decay, it must itself be radioactive, and when

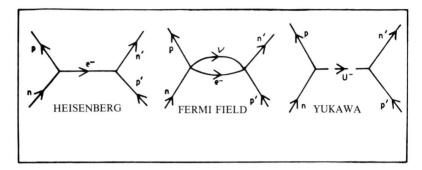

Figure 1. Models of the strong nuclear charge-exchange force. p, p′, protons; n, n′, neutrons; e⁻, electron; ν, neutrino; U⁻, heavy quantum.

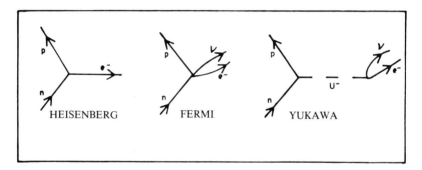

Figure 2. Models of the (weak) β decay interactions. (Notation as in Figure 1.)

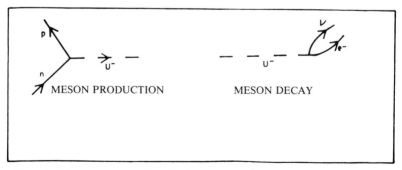

Figure 3. Other processes predicted by Yukawa's theory. (Notation as in Figure 1.)

it is produced in the cosmic rays, it must decay in a small fraction of a second. This was pointed out later by the Indian physicist H. J. Bhabha,[12] and estimates of the half-life were then made, but not until 1938.

I emphasize that Yukawa introduced two new coupling constants or "charges", having the same physical dimensionality as that of electric charge, but of very different strength. The larger one measures the strength of coupling of the meson to the heavy particles; it accounts for the exchange of mesons that gives rise to the binding of nucleons to form nuclei, for the scattering of nucleons on each other, and for the production of free mesons when sufficient energy is available. The other charge is much weaker (about 10^{-8} times smaller than the strong coupling constant). It measures the coupling strength of the meson to the light particles, electron and neutrino, and accounts for the weak interactions: nuclear beta decay, free neutron beta decay, and meson decay. Heisenberg's and Fermi's theories each involve only one coupling constant but Yukawa's theory is in much better accord with nature.

We quoted Yukawa earlier on the usage of the term "elementary particle" at the beginning of the 1930's; it meant simply electron or proton, and there was a prohibition of new elementary objects that was sometimes unspoken, at other times explicit. When Pauli proposed a new type of particle, the neutrino, he waited three years to publish his idea — and then did so only in the form of a discussion remark. When Dirac was "led to a new kind of particle caused by a hole in the distribution of negative-energy states," and the symmetry properties of the theory suggested that the particles have positive charge and electronic mass, Dirac still called it a proton.

I just didn't dare to postulate a new particle at that stage because the whole climate of opinion at that time was against new particles.[13]

These are some of the reasons why Western physicists did not anticipate Yukawa's nuclear force model. When the "heavy electron" or "mesoton" (as the muon was then called) was found, Leon Brillouin identified it as Yukawa's U-quantum. Referring to Heisenberg's suggestion to use the Fermi β-decay mechanism as a realization of the charge exchange nuclear force, Brillouin said:

> A Japanese, Yukawa found in 1935 a very simple way of improving the theory. "Fermi's Field" surrounding a heavy particle corresponds to the exchange of an electron with a neutrino. Yukawa had the idea of admitting that this field should actually correspond to the exchange of one new particle differing both from an electron and a neutrino.[14]

On occasion the scientific process is described by political analogies; sometimes the analogy is mechanical or thermodynamic. Such analyses seem inappropriate to apply to Yukawa's achievement, for they emphasize a shift in the understanding of a *given system*. But Yukawa's prophetically entitled "On the Interaction of Elementary Particles. I." is a door opening on a world of high energy processes, involving the creation and annihilation of new ephemeral substances (the mesons, unstable leptons, strange and charmed particles, etc.) of astonishing novelty. One is justified in regarding this as a lesser revolution than, say, quantum theory or relativity, only by placing an exaggerated value on modes of thought, as opposed to understanding the *content* of nature.

Although Yukawa has his own unique philosophical orientation, it would be inappropriate to discuss Yukawa's thought without mentioning the philosophical ideas of Sakata and Taketani, as it would be to discuss Einstein's early thought without mentioning Ernst Mach. To compare the two cases may appear absurd; after all, Sakata and Taketani were *students* of Yukawa, while Mach was almost forty years older than Einstein. Nevertheless, the relationship between Yukawa, and his brilliant students was a complex and reciprocal one, and was especially important in the social and political circumstances of Japanese physics in the 1930's.

Between the first meson paper and Yukawa's "short note" identifying the cosmic ray "heavy electron" with his U-quantum, Yukawa published seven papers, five of them jointly with Sakata. Sakata wrote that Taketani used to visit the Osaka group after his graduation from Kyoto University in 1934. "He always had ideas about the theory of the nucleus and we used to discuss it until late at night. Sometimes he talked about Hegel's logic and Yukawa would listen with great interest." Taketani's graduation thesis at Kyoto had been a study of the logical structure of quantum mechanics, which served as the starting point for his "three-stage methodology" of physics. This adds a third "substantialistic" stage between the "phenomenological" and "fundamental" stages that are often assumed. In the substantialistic stage one identifies the substances in the model: for example, electrons and the nucleus are identified as the constituents of the Bohr-Rutherford atom. Taketani's methodology is materialist, emphasizing concrete objects, and it is dialectical, since the final stage of one theoretical development is the beginning stage of the next.[15]

Although Sakata was the more orthodox Marxist, Taketani was active politically and wrote for a radical journal called *World Culture* (*Sekai bunka*) that served as an intellectual spearhead of the movement against fascism. In 1938 there was a crackdown, and Taketani was arrested. He wrote about this:

> The unlawful acts for which I was held were: my analyses of quantum mechanics, my analyses of the development of nuclear physics, and my methodological approach to the meson theory — in short, my research activities on natural dialectics. I was forced to state that I had participated, through my research, in the cultural movement of the popular front under instructions of the Comintern, thus helping to promote the Communist Party in Japan.[16]

Sakata and Taketani do not claim to have influenced Yukawa's meson prediction, except most indirectly. Sakata wrote, "Yukawa is really not a dialectical materialist." Yukawa rarely referred explicitly to Taketani's methodology. However, in a preface to a recent biography by the poet Sosuke Takauchi, Yukawa discussed the divergence problems of quantum field theory that he regarded as more fundamental than the theory of the nuclear force.[17] Using the language of Taketani, he asserted that his main intellectual interest lay in the "essentialistic" (i.e., fundamental) stage of the theory. In the future, he said, there must be fusing together of the last two stages, the substantialistic and the essentialistic. Quoting the Chinese poet Li Po ("The universe is an inn that accomodates all things; time is a traveller of one hundred generations."), he said that one day in 1966, he conceived of the idea of the "elementary domain", a modification of the ideas of space-time and causality — requiring the use of difference equations, rather than differential equations, an idea recalling Heisenberg.

Yukawa did not identify himself with the Sakata-Taketani methodology or with their political philosophy, but his close association with them in the 1930's helped to relieve his sense of scientific isolation, while strengthening his independence of the various "authorities" — among them the state and its institutions, the family and its traditions, and intellectual authorities, both at home and abroad. In formulating the meson theory, Yukawa, as the novelist Soseki had foreshadowed, showed his intellectual independence of the old Japan *and* the West.

This introduction to *Tabibito* has dealt, like the autobiography itself, with Yukawa's life and work to 1935, the year in which his meson theory was proposed. The English-speaking reader can sample Yukawa's essays on science and social issues in *Creativity and Intuition; A Physicist Looks at East and West* (Kodansha, 1973), translated by John Bester. Yukawa's *Scientific Works* (including a selection of English translations of Yukawa's Japanese writings and all of those in Western languages), has been edited by Yasutaka Tanikawa, who furnishes a biographical sketch (Iwanami Publishers, 1979).

I conclude with a brief mention of high points of Yukawa's career after 1935. Until about 1948, Yukawa and his collaborators worked on problems of meson theory and nuclear forces, but Yukawa became increasingly occupied with fundamental aspects of the theory of elementary particles. After 1948, in which year he became a member of the Institute for Advanced Study in Princeton, most of Yukawa's work concerned non-local field theory and extended models of elementary particles.

Yukawa moved back from Osaka to Kyoto in 1939, assuming a professorship there. In the same year he made his first trip abroad, cut short by the outbreak of war in Europe. In 1940 he was awarded the Imperial Prize of the Japan Academy and in 1943 the Cultural Medal, Japan's highest award. After a year in Princeton after the war, he served from 1949–1953 as a professor at Columbia University. He was awarded the Nobel Prize for Physics in 1949.

A grateful Japanese government honored him by establishing the Research Institute for Fundamental Physics in Kyoto in 1953. Yukawa was its first director, a position he held until his retirement in 1970. He had founded, also in Kyoto, the international journal *Progress of Theoretical Physics* and he remained its editor until his death on September 8 1981.

REFERENCES

1. Kenkichiro Koizumi, "The Emergence of Japan's First Physicists: 1868–1900," *Historical Studies in the Physical Sciences 6* (1975), 3–108, on pp. 82–94.
2. Natsume Soseki, *Sanshiro*, English translation by Jay Rubin (University of Tokyo Press, 1977).
3. *Creativity and Intuition*, a collection of Yukawa's essays in English, translation by John Bester (Kodansha International, Tokyo, 1973).

4. H. Yukawa, *Shizen (Nature)*, July, 1975, pp. 28–39; *Special Issue on Forty Years of Meson Theory*.

5. Cf. Joan Bromberg "The Impact of the Neutron: Bohr and Heisenberg," *Historical Studies in the Physical Sciences 3*, (1971), 307–341.

6. Gamow, pp. 55 and 56; emphasis supplied by Gamow.

7. L. M. Brown, "The idea of the neutrino," *Physics Today*, Sept. 1978, 23–28.

8. W. Heisenberg, "Über den Bau der Atomkern," *Zeit. f. Physik 77* (1932), 1–11; *78* (1932), 156–164; *80* (1933), 587–596. (Part I and part of Part III are given in English translation in D. M. Brink, *Nuclear Forces* (Oxford, 1965), 144–160).

9. E. Fermi, "Tentativo di una teoria dell ' emissione dei raggi 'beta'," *La Ricerca Scientifica 4(2)* (1933), 491–495; "Tentativo di una teoria dei raggi β," *Nuovo Cimento 2* (1934), 1–19; "Versuch einer theoria der β-Strahlen.I.," *Zeit. f. Phys. 88*, (1934) 161–171. (The last paper is given in English translation in Charles Strachan, *The Theory of Beta Decay* (Pergamon Press, 1969), 107–128.)

10. Ig. Tamm, *Nature 133*, 981 (1934); D. Iwanenko, *ibid.*, 981.

11. Hideki, Yukawa, "On the Interaction of Elementary Particles. I.," *Proc. Phys.-Math. Soc. Japan 17* (1935), 48–57.

12. H. J. Bhabha, "Nuclear Forces, Heavy Electrons and the β-Decay," *Nature 141* (1938), 117–118; H. Yukawa and S. Sakata, "The Mass and Life Time of the Mesotron," *Proc. Phys.-Math. Soc. Japan 21* (1939), 138–140.

13. P.A.M. Dirac, "The Prediction of Antimatter," *The 1st H. R. Crane Lecture* April 17, 1978 at University of Michigan, Ann Arbor.

14. Leon Brillouin, "Individuality of elementary particles — Quantum statistics. Pauli's principle," in *New Theories in Physics*, Conference held in Warsaw, May 30 to June 3, 1938 (International Institute of Intellectual Co-operation, Paris, 1939), pp. 119–172.

15. S. Sakata, "On Interpretation of the Quantum Mechanics," in *Supp. Prog. Theor. Phys. No. 50* (1971), 171-184.

16. "Methodological Approaches in the Development of the Meson Theory of Yukawa in Japan," in *Science and Society in Modern Japan*, ed. S. Nakayama, D. L. Swain, and E. Yagi, 1974 (MIT Press, Cambridge, Mass.). Taketani's article is adapted from *Supplement of the Progress of Theoretical Physics No. 50* (1979), 18-24.

17. Sosuke Takauchi, *Order and Chaos* (Kosakusha, 1974) in Japanese; enlarged and revised edition in 1979.

Chapter 1

The Home of Knowledge

Many people think that Kyoto is my hometown. It is actually the place where I received most of my education. After graduating from the University in Kyoto, I lived for a time in the vicinity of Osaka, but then I returned. When I came back from America in 1957, I did not feel that I was "home" until the train approached Kyoto through the Tōyama tunnel.

But, unquestionably, I was born in Tokyo (at Nibanchi, Ichibei-cho, Azabu). I cannot remember the house of my birth, not even the plum blossoms there. I know them only through my mother, but in my mind I know they are extremely beautiful. (Perhaps my subconscious tries to beautify my birth.)

I lived in my house at Azabu for fourteen months. Then my father, Takuji Ogawa, who was the head of a geology research center in Tokyo, accepted a professorship at Kyoto Imperial University and moved there with his family. He taught the first geography course at Kyoto University.

Azabu was burned during the War and is no longer recognizable as it was. Our house was at the end of a narrow path on a slope. One of our neighbors was Takuma Dan, and Kaneko Yanagi had a large and impressive house there (I am told).

When I was a small boy in Kyoto, I used to beg my mother to tell me about the house and she would begin: "It was not a large house, but it was sunny and comfortable. You were born on a cold day and ... ." Of course it was cold, it was the 23rd of January! The plum blossoms must have been still hard buds. My oldest brother, Yoshiki, was six; my second brother, Shigeki, was four. They barely remember the house, but my sisters Kayoko and Taēko, who are older, seem to remember it better. Fifty-year-old memories are probably not very accurate.

After the luggage and parcels were shipped from Tokyo to Kyoto, the family stayed at an inn at Shinbashi, near the railway station. "The train tracks at night emit light, and I still remember the blue glow," my brother says. We moved in late March; the nights must have been cold.

The year was 1908, not long after the Russo-Japanese War. The atmosphere in the country was probably tense, and the vicinity of Shinbashi Station must have been incredibly dark by modern standards. Under the oil lamp that hung from the low ceiling of the old inn where we stayed, I can imagine how the family members harbored in their hearts the hopes and fears of a new beginning. (But the inn may already have had a dim electric light.)

My own memories begin from the time we moved to Kyoto, so perhaps Kyoto *is* my hometown, after all. My earliest recollection is the image of myself on my mother's back. I believe we were standing on the bridge that connected the railroad platform with the main building of Kyoto Station; I was half asleep. It was probably not the time of our arrival in Kyoto that I recall, because the image is much too vivid. I can see the dirty ceiling and the blackened windows of the covered bridge, and I can hear the steam whistle of the train.

There is another similar memory: I was being carried in a garden on someone's back, probably one of the maids who served our large family. I can hear a sleepy lullaby and I am drowsy. The garden is covered with moss, and the white wall of the storehouse is bright with the thin rays of the winter sun. This memory seems to refer to the house in Somedono that I will describe later.

When our family arrived in Kyoto, we found that no house was available. We were supposed to rent a section of Engakuji Temple in Yanagifuro, close to the old Imperial Palace, but it was not ready for us. For a time, we settled in Sawabun Inn at Sanjo Fuya, where we had a very large room. The unsettled living in the Inn must have been hard on my mother. Father had preparations to make for his new position. The two pre-school boys played games beside the desk where Father was working. The baby, myself, must have been crying from time to time.

During this hectic time, a misfortune occurred. One night, Takuji, my father, began to run a high fever and complained that his arm hurt. At first my mother thought he was tired from carrying heavy luggage during the long trip. But his face became red with fever, and he said the pain was unusual. The arm hurt from shoulder to wrist and it was swollen. A doctor came from Kyoto University Hospital and the next day my father entered the hospital. My father was cured by an operation under full anesthesia; fortunately, no mistakes were made.

While my father was in the hospital, the family moved into the Engakuji Temple. It was hard for my mother to make the move with so many children. My parents had no relatives living in Kyoto, although both of them came from Kishu, not far away. Yanagifuro, where the temple is located, is still a quiet district; in those days it was at the lonely edge of town. In that year, only twelve years after the establishment of Kyoto Imperial University, the population of Kyoto numbered just over 300,000. We lived in the temple for about one year, and we changed houses fairly often after that, sometimes for the convenience of the landlord, but more often for our own reasons.

My father's mother and my mother's parents were living with

us, my father having been adopted upon his marriage. After myself, my two younger brothers Tamaki and Masuki, were born. It was a large family — and on top of that, my father's library was expanding rapidly!

My father, although a specialist in geology and geography, had wide interests and bought many books about other fields. For example, he was very fond of art. When interested in something, he could not rest until he had collected all possible writings on that subject. For instance, when he took up the game of *go,* he bought all the books on it that he could find. They would fill up the library and then spill into the study. When all storage places were full, my father would say to my mother, "We shall have to move again. Where can we find a larger house?" In those days, university professors must have been much wealthier than they are today!

My father rode to the University each day in a rickshaw and returned the same way. These hand-carts did not have rubber tires, and I can still remember the loud noise made by their wheels. But the family was not always well-off financially, because of the many books and art objects my father bought. Sometimes he doubted whether he could afford to send all his sons to the University to study. Without my mother's strong urging, I might not have received university training, and I would not exist today as I am.

Although I do not believe in predestination, I cannot avoid a strange feeling when I think about this. A man cannot know in advance what will shape his career; I could not have known that an idea would come to me twenty years later that would make a contribution to physics, and that I would influence the shape of physics.

We moved finally to a house in Teramachi (Hirokoji Noboru, Somedono), just north of the Nashiki Shrine. The house had belonged to a nobleman connected with the household of the Emperor. Its garden was covered with moss. Our landlord, the aforementioned nobleman, was named Rokujō. I never saw him, but according to my older sister, he had a squarish face that was fierce and frightening to a child. Everyone called him "Deaf Rokujō-san."

Where did this name come from? My older sister says that she can still remember him riding a horse, splendidly costumed during the annual Aoi Festival in Kyoto. I always thought he was really deaf, but my sister later told me that he had no physical defect in this regard, but only *pretended* not to hear when he was told something he did not like, so that the neighbors jokingly called him "deaf." In those days, Kyoto had many noblemen with high ranks and thin wallets, and just as many with eccentric reputations — "Deaf Rokujō-san" was one of these.

The house was on Teramachi Street. The rooms at the back of the house were large and old-fashioned; my grandparents had their rooms on the second floor. To the left of the gate, set against the wall that surrounded our property, was a room with a high floor and a lattice window. I played there with my brother Shigeki, and we looked out on the street through the window. Even today, Teramachi is a narrow street. In those days a small street railway line ran on one side of the street, one of the oldest such lines in Kyoto. Today (in 1957), the street car runs on Kawaramachi Street, east of Teramachi, and the tracks have been removed from Teramachi.[1]

On the other side of the street almost facing our house, was the main gate of the temple Shōjōkein, the headquarters of the Jodo sect of Buddhism. For some reason, we called this "Jōken Temple," and my brothers often played on its grounds. As one entered its gate, the main temple building stood to the left. At the roof, one could see the chrysanthemum seal. Passing through the elevated corridor that connected the main building

---

[1](In 1979, all the tram tracks were removed from Kyoto, *trans.*)

with the office and the priests' living quarters, one looked over the cemetery. My brothers played hide-and-seek there, and I also sometimes played there, although I remember almost nothing about it. Wait, there is one thing —

My eldest brother, Yoshiki, was mature for his age and the next oldest, Shigeki, was full of confidence; I, the third son, was already conscious of pressure from the older boys. As I was running through the graveyard of "Jōken Temple," I slipped and struck my head against a stone grave marker; I saw darkness for a moment. I started to cry, but my brothers were already far away. As I lay on my back, the sunbeams that shone through the leaves of the cherry trees hit my eyes, and I gasped: they were like countless stars — the midday stars! Much later, when the idea of the meson occurred to me, I caught a vaguely remembered glimpse of those midday stars.

My father, Takuji, traveled all over Japan doing geological surveys when he was a young man. At thirty years of age, he was appointed a delegate to an International Geological Conference in Paris. He was always intent on his work. To a child, that sort of scientist is sometimes too strict a father; at times he seemed insufficiently concerned about his child.

In later years, after I had been adopted by the Yukawa family, I used to return to visit my father. We would talk for hours, sometimes about our researches, sometimes about current issues. It was then that I discovered that he had another side. But when I was a child, he seemed to me a humorless man. Often he would say to my mother, "One shouldn't spoil children." And she would answer, "But they are still so small." It does seem as if my father did not approve of childlike behavior in children!

I do not recall ever being hugged by my father. Neither did I ever ask him for toys. Perhaps he expected a child to have a man's outlook. When I entered the Third High School in Kyoto (junior college level), the Headmaster, Sotosaburo Mori, said in his welcoming speech to my class, "From today on, I will treat you as gentlemen." Mori was speaking to boys of seventeen and eighteen, and his speeches were famous in Japan. Perhaps my father's attitude was a similar one.

Toward my brothers, his attitude was the same; I do not think any of them had ever been embraced by my father. His authority was absolute, and it was my mother's duty to support it. She gave up her own personal life to take care of the children and the old people in the house. Mother might have done her own shopping once a month, but she always returned home as soon as she was finished. She did read a great deal, but that was also done for the children.

By today's standards, Mother was very old-fashioned and there is no doubt that she accepted her household tasks as her obligation. Her only joy, it seemed, was in raising the children. She died in 1943, and had never seen a movie in her life. The very idea of pleasure seemed not to exist for her. As a child, I always felt sorry for her, and I still wonder if she was really satisfied with her life.

As my mother was always busy, I grew close to my grandmother, my father's mother. Her name was Mie Asai. She was of an open and outspoken disposition and used to take me to visit the numerous Buddhist temples and Shinto shrines of Kyoto: Kyoto, seen from the stage of Kiyomizu; the autumn leaves of Tōfukuji Temple; Chion-in's huge roof and its roof-tiles: The buildings and the natural beauty of Kyoto have made strong impressions on me.

Both my parents walked a path without lifting their eyes from it. Perhaps that is the heredity that led me to walk the path of science; yet, as a child, I was not fond of science. Although I received good grades in elementary school, I was far from a genius in middle school or in high school. My eldest brother, Yoshiki, likes to say jokingly, "Hidé was not a bright child; he was stubborn, and gave us trouble." But I became very absorbed in my activities. When I was given building blocks to play with, I would do so by myself for hours at a stretch. Our house had a sunny veranda facing the garden which had bushes and stone lanterns, and I would play with the blocks there — building houses, gates, and towers.

How did this look to others? My father's mother would bring me snacks. "Are you still playing with those blocks? What a busy child!" she would say in a kindly voice. As far as I could tell, the towers I built were as gorgeous as the Tower of Yasaka, and my houses as impressive as those of the Imperial Palace. "Grandma, I'll make you the Honganji Temple." I would take the tower apart and start the new task. When it was finished, my grandmother would say, "What a beautiful temple! I'll have to make a pilgrimage to it." Then she would pretend to pray.

One day Grandmother bought me a jigsaw puzzle of twelve pieces, with a brightly colored picture. I played the game of assembling it with great concentration, but after several times it became too easy. Nothing could have been easier once the completed picture was memorized, and my curiosity quickly faded as I noted, without effort, the shape of each piece. "Look, Grandma, I can do the puzzle face down!" I could not see any picture facing me but I put the pieces together, and when I flipped it over, there was the picture. "Well," said my grandmother, "*there* is a bright child — perhaps the brightest in the family." My grandmother thought highly of my capabilities, a feeling not shared by the rest of the family.

That grandmother died in the house at Somedono, and so I was lonely when we moved to another house, at Higashizakura. But I became the favorite of my grandparents on my mother's side. The new house was owned by a viscount named Keishi Toyōka. The county hospital was nearby, and so was Kuninomiya Palace. Beyond the Palace flowed the Kamo River. Once a year, Rokusainenbutsu (a dancing incantation of Buddha) was performed inside the palace grounds, and on that day the public was admitted. A large go-board was placed in front of the gate, and on it a lion dance was performed. The bright orange and green of the lion's costume, and its fierce-looking mask, impressed me deeply.

The gate of our house was like that of a temple; its roof-tiles were shaped like peaches. Inside the gate there was a shelter for the servants of visitors. After this shelter, there came the storage house, and then the main entrance, with a stoop in front. It was the old-fashioned house of a nobleman. I remember that several bamboo trees stood to the left of the gate, and a holly tree stood to its right.

As one passed through the wooden gate to approach the house, there was a courtyard to the right, in one corner of which stood a small shrine. Next to it was a detached room, which became the living room of my grandfather, Komakitsu. In the courtyard between this room and the main house, my grandfather cultivated morning glories and chrysanthemums. Once he took me to a morning glory competition. He also used to take me to see sumō wrestling.

The town is all changed now and it is hard to describe the exact location of the morning glory contest, which was held in the garden of a large house near the northeastern edge of Hanamikoji, several blocks from the Kamo River, along Shijō Street. There I saw, in countless numbers, flowers as large as my face. There were also very intricate flowers whose petals trailed like threads. The sumō wrestling was held on the grounds of Kennin Temple, and I still recall that one of the wrestlers, named Tachiyama, was much stronger than the others. The place where the wrestling was held is also gone now.

From time to time, small shows took place on the Kamo River banks and booths were set up there. During such times part of the bridge railing was removed and steps were built, leading down to the river bank. I never bought anything at those booths, but my heart leaped up with excitement as I strolled past them along the river.

In our house it was the duty of the boys to clean the chimneys of the oil lamps. One cold winter day, I heard a strange piping sound as I sat on the wide porch in front of the storage house, cleaning a lamp chimney. It was a simple and warm sound, full of melancholy. I asked my grandfather, "What kind of instrument is that?" "It is the reed-pipe, used in court music," he answered. The music came from the house of our landlord, Toyōka-san. Though ignorant of court music, I was drawn to it in a strange way. Even now, I feel nostalgia at the sound of the music that is played at Shinto weddings.

I do not remember Viscount Toyōka at all; he was not much in evidence. Later, I found out that he was a member of the House of Lords, for whose meetings he traveled often to Tokyo. He organized The Court Music Society in Kyoto. In any case, here was a man who spent days playing the reed-pipe: a tale from a more relaxed age. He owned a large library of court music which, regretfully, was burned in the last war.

I have never had many close friends. That may be partially because of my personality, but it is also typical of the character of Kyoto, where I spent my early years. The houses of Kyoto are designed to separate their inhabitants from the world outside. Even in the crowded shop districts, only the stores front the street. When you pass into the living quarters, through the curtain that hangs at the back of the shop, there lurks a quiet darkness that is cut off from the outside. There are even shops whose mechandise is not visible from the outside.

In the suburban areas, high white walls continue without break. There is the high gate with the heavy roof; there is the shrubbery; the house is always at the back. Although there are gardens full of sunshine, one would never know from the outside. Those structures are ideally suited to breed the kind of character one finds in Kyoto. No, perhaps it is rather the product of Kyoto people, who so easily close their hearts to those outside.

I never played with children of my own age, although some must have lived in the neighborhood. And I never felt any curiosity about the other houses. I felt that existing in a small limited world was natural. Such confined circumstances can produce in a child a rich imagination and a romantic temperament. My oldest sister had finished her schooling and was at home, but she was too old to play with me. During the day, until my brothers returned from school, I was alone with my younger brother, who was too small for me. Sometimes I would leave the house and walk aimlessly.

Honzenji Temple had a Laughing Emma (a guardian of Hell). It occupied a small dark house, and if you looked at it steadily for a while, it seemed to laugh at you. I went there often, although it frightened me. Nashiki Shrine is beautiful and green; just past it is one of the gates of the Imperial Palace. Inside the gate there extends a straight road made of cobblestones, having a white wall on its right and trees on the left. A large lotus tree bore small pale-yellow flowers in spring. In the autumn, it was loaded with small scarlet fruits. Beetles lived in that tree. I stood under its leaves looking up for them; but one cannot find these beetles during the daytime.

My older brother caught the beetles for me. In the early morning, even before washing, my brother and a friend ran to the gates. A solitary arc lamp stood there. The beetles, attracted to the lamp from the lotus tree during the night, are tired by morning and fall to the ground, lying quietly in the dewy grass. Their black wings gleam.

The beetles with one horn that divides into two at the tip we called "Kabuto." The ones with two horns were "Gengi." Those with no horns were "Bōzu." I kept them all in a wooden box, and fed them sugar water. Sometimes I would take them out and let them wrestle, or pull paper carts. I made air holes in the lid of the box, and I weighed it with a stone so that my beetles would not escape during the night.

Hideki and his brothers at their home (Higashi-sakura-machi, Kyoto). From left:
Shigeki (later adopted into the Kaizuka family, Professor Emeritus of Kyoto
University, Chinese historian), Hideki, Tamaki (Professor Emeritus of Kyoto
University, scholar of Chinese literature) and Yoshiki (deceased, Professor of
Metallurgy, University of Tokyo).

Hideki with his father, Takuji, his mother, Koyuki and brothers. From left in the back row: Tamaki, Shigeki, Hideki and Masuki (died in war).

Chapter 2

My Father

I was in Japan for a short time in the summer of 1949, between trips to the United States, and in that particular summer there was a reunion of my former elementary school class attended by about twenty of us. That so many were present after thirty years was surprising, as there were fewer than one hundred in my class. That may also have to do with the city of Kyoto, where many people tend to carry on family trades, even today. Most Kyoto women marry in or near the city; and, of course, Kyoto was not damaged in the war.

All my classmates were in their forties; some I could still recognize, while others I could not remember. But after we talked, cloudy memories began to return. Because interests were now so different, our conversations tended to leave the present and fly to the past, where all of us were on common ground.

"I never thought Yukawa-san would become a physicist," said Mrs Mikiko Narikawa, who sat next to me in first and second grades. "I thought that you would go into the literary field, although I never gave the matter much thought." Some others agreed with her; perhaps that is how my friends saw me as a boy. "Yes," I answered, "I read a lot of literature as a boy," and I was reminded of my father's schooldays.

My father, Takuji, had also been inclined toward a literary career until he was halfway through the First High School (in Tokyo). If it had not been for a chance event, he would probably not have become a geologist; he had intended to study philosophy. He read English novels in the original, and not only in order to help learn the language. One of his favorite authors was Dickens.

For a time, my father belonged to a literary club (Garakuta-bunko). That was in the first part of the twentieth year of Meiji (1897). Sanji Tokai's *Strange Fortunes of a Beautiful Woman* and Soho Tokutomi's *The Young Generation of the New Japan*, were being eagerly read by the younger generation. *Student Spirit,* by Shoyo Tsubouchi, was being discussed. *Kenyusha,* with Kōyō Ozaki as its driving force, was finally beginning to spread a fresh atmosphere.

My father lived near Kōyō's house, at Handa, while attending school (in Tokyo), and he became a member of the "bunko."[1] Sometimes Kōyō brought a circular letter to my father; sometimes my father went to Kōyō's house to get one. I am told that often they held discussions in Kōyō's study. Let me write about my father in more detail.

My father, Takuji Ogawa, was born at Nanki, in the third year of Meiji (1870). He was the second oldest of the sons of Nammei Asai, a teacher of Confucianism for Tanabé-han.[2] Until he was adopted upon marriage, my father had the name Asai, Nammei taught Chinese learning to the children of the *han*, but after the *hans* (and the *samurai* class) were abolished, he opened a private school in a small village. Although my father entered Wakayama Junior High School at the age of fourteen, he had already been taught the Chinese classics (in the Japanese language) by his father.

[1]fellowship
[2]han: feudal clan or its territory

When my father was about to move up to the third level, an amendment to the school system was decreed, combining the second and third levels into one. This disappointed him enormously, the more so since he was already late in entering school. Also, some other students, who had been studying in Tokyo, came home on vacation and gave parties that he attended. Listening to their stories, he developed a longing for Tokyo and the New Knowledge to be had there.

In the nineteenth year of Meiji, at the age of seventeen, Tokuji went to Tokyo. He was not a wealthy student. Temporarily, he enrolled at the Tokyo School of English, which was a famous preparatory school. The next year, he applied for admission to the Naval Academy, planning to have his education paid for with his officer's salary. But fortunately (I think I may say so) he failed the physical examination. The fact that he had an academic, rather than a military career, certainly influenced his children.

He had not decided on an academic major when he entered the First High School, which was one reason he became close to Kōyō Ozaki. My father always felt a great nostalgia for this period of his life. In later years, when he talked with his children about literature, he would refer to Kōyō with a special feeling; however, he showed little interest in modern literature after the novelists Ogai and Sōseki.

After two years in Tokyo, he was adopted into the Ogawa family. Up to that time, Takuji's brother had paid for his education, but the brother, a lower state official, could not afford to support him any longer. Someone who knew of my father's financial plight had arranged for the adoption, which was to be followed later by marriage into the family. His future father-in-law was Komakitsu Ogawa.

Komakitsu was also from Kishu and he had participated in the skirmishes that had occurred in the final years of the Tokugawa Shogunate. Afterwards, he studied under Fukuzawa at Keiōgijuku (now Keiō) University. He became the principal of a school, and later he was employed by Yokohama-Shikin Bank. Sometimes he taught classes at Keiō. Interestingly, he too had been adopted, his original name being Nagaya. One of his friends, Nobukichi Koizumi, was the father of Nobuzo Koizumi, who later became the president of Keiō University. Seeing Takuji undecided about his future, his father-in-law must have said to him, "Go talk to Koizumi-san. He may give you an idea." Koizumi possessed a vast knowledge of Occidental science and had a special interest in applied science and technology. Whenever Takuji talked with Koizumi, it whetted his interest in science. There were at least two definite reasons why my father chose geology as his specialty.

In the spring of 1891, he went to visit his mother, who was ill, and upon his return to Tokyo he came down with a bad case of influenza. It was just before an examination period, but he took the exams in spite of his illness, which was weakening his body. After his recovery from influenza, he was continually plagued with insomnia, which the doctor diagnosed as due to nervous tension. My father was told that walking was the best cure for insomnia, and he took long walks in the suburbs on weekends. But it was to no avail, and finally he decided to skip the next set of examinations and escape the heat of summer by going to a mountainous region, Gotenba, with a friend. They obtained a room at a temple of the Shin sect of Buddhism. My father continued to read novels in English, while he marvelled at the beauty of Mount Fuji. "I want to conquer that mountain someday," he said. At his friend's reply: "Not in your state of health," the desire to walk up the mountain rose more strongly in his heart.

My father returned to Tokyo in September and took the make-up examinations, moving up to the second level. But as his insomnia remained, he decided he would discontinue his education for the time being and travel to his home area of Kishu. He was staying with his adopted family in Yokohama in October, and preparing for the trip when the famous Nōbi earthquake struck. It was October 28. A strong shock was felt in Yokohama in the morning, followed by a slower shaking that continued for some time afterward. The newspapers put out extra editions, and by the next day the huge damage done by the earthquake became known.

Although common sense should have dictated that the planned trip be abandoned, my father chose, instead, to leave immediately. "But the railroads are not functioning in many places." objected the family. "Then I shall walk." He planned to observe the effects of the earthquake on foot, if necessary. Three days after the quake, on the last day of October, he left. Taking the crowded train, he met an older student from his school, Tetsugoro Wakimizu, who was accompanying a professor to do research on the effects of the quake in Ōita. When my father explained his own plans, Wakimizu said: "That is not an easy trip. It is not in your area of study, and there is no need to expose yourself to the danger." That was his advice, which my father disregarded.

He got off the train at Nagoya to find the city in a state of confusion surpassing imagination. There were burnt and collapsed houses. People slept outdoors for fear of a recurrence, and it was cold for November. My father sympathized with the people, but felt that he could not stay there. Fortunately, he found a rickshaw that had come from Ōita, and he took it westward. He wanted to see the extent of the damage, driven by his desire to learn. He decided there and then that he would bend his efforts to oppose the might of nature.

It was horrifying! The train rails had fallen beneath the elevated roadbed. Some people were still removing possessions from half-fallen houses. The huge cracks in the roads and the dikes astonished my father, who was seeing for the first time, an area devastated by an earthquake. He noted how the strong destructive forces were applied from below when he came across a temple's bell-house standing on its side and retaining its form, although its foundation stone was ripped apart. He was astounded, perhaps even inspired, by the power of nature, even though he sympathized with its unfortunate victims. This trip motivated him to study geology.

As he approached the Ōita train station, things gradually became calmer. The station was crowded with people, but he boarded the train for Osaka. He arrived next day at Wakayama, where he stayed at the Nagaya household. He intended mainly to rest. Borrowing a rifle from his uncle, he went into the forest of Wakayama Castle to shoot birds, but without success. "You can't shoot fowl while walking in broad daylight," said his uncle, laughing. "You must, instead, surprise the bird in its nest at dawn or twilight." My father learned from these words, and later, as a geologist, when he went to collect geological specimens, he used to try to strike at the "nest" of the minerals.

My father stayed with the Nagayas for about a month and left at the beginning of December, planning to travel around Nanki. He observed the effects of a mudslide that had occurred during the Totsugawa flood the year before; he came to Torohacho, along the main Kumano River. The whole trip made him aware of the awesome size and power of natural forces. He wrote four Chinese poems on the theme, "Looking out over the Pacific Ocean from Shinozaki." He was well-versed in Chinese.

This trip made up my father's mind. The earthquake of Nobi, the wilds of Nanki, and the complex forms of its seashores awakened his urge to know more. Having come to a decision, he returned to Yokohama as quickly as possible. There he discussed the matter with his adopted father, and then went back again to Tokyo. In the next year, my father legally took the name Ogawa and was transferred to a geology program. It was then that his life began to center around geology. He traveled all over the country doing research until he left the Geological Research Center, but his passion to learn lasted all his life.

Takéo Kuwabara, who was a classmate of my brother, Shigeki, during high school, became a prominent authority on

French literature. I am told that Kuwabara-san still says: "I was always afraid of Takuji Ogawa." Indeed, he was once reprimanded by my father. Since the anecdote sheds light on my father's character, I will include it here.

Kuwabara was a member of the mountaineering club at the high school. On one occasion, while climbing in the Japanese Alps, he lost contact with his group. The whole school was worried, and his friends organized a rescue party. It turned out that Kuwabara had injured himself, and had to be brought down to the town of Fuji on the back of a guide. Before returning to Kyoto, an operation had been necessary. His friends were anxious to hear his story, and he probably felt like a young hero. After recovering from his wound, Kuwabara came to visit my brother Shigeki. The two were talking in one of the rooms; Kuwabara's voice sounding young and loud. My father, sitting in his study, noticed his cheerful tone and summoned him. Although Kuwabara-san was a vigorous young man, he was still just a high school student while my father was a professor at the University, so he probably entered the study humbly and begged pardon for all the anxiety his adventure had caused. But my father rebuked him, saying, "Don't climb mountains only to risk your life."

Kuwabara tried to explain the circumstances, but my father would not listen to him. He delivered a long lecture and sermon. My father loved youthful spirit, which is why he continued teaching until his retirement, but his genuine love sometimes came in the form of strong words. He spent more than an hour to make Kuwabara realize the importance of life.

My father might have meant that one must have a good reason to climb mountains. He, himself, had climbed most of the mountains of Japan, and he drew upon his experience. The reason need not be scientific; it can be for sport, but there must be a purpose. The idea of conquering a mountain "because it is there" was foreign to my father, and to the Japanese society in general. During his geological surveys, he had sometimes to put

himself in danger. He was not able to join the expedition that
was at Mount Azuma when it erupted. If he had, he would have
taken a dangerous chance; perhaps it was this that made him
rebuke Kuwabara all the more.

The eruption of Mount Azuma in the twenty-sixth year of
Meiji was an important event in the history of Japanese geology. The first eruption occurred in May, and a second in June.
The Department of Agriculture and Commerce decided to send
engineers, and the Geology Department of Tokyo University
was to send its representatives with the expedition. My father
wanted to go, but his father-in-law, Komikitsu, was ill, and as
the end-of-term examinations were close, he decided against it.
After visiting Komakitsu in Yokohama, and returning to
Tokyo, he read in the newspapers that two members of the
expedition had been killed by the volcano eruption. Those two
had proceeded ahead of the others and a third eruption had
occurred as they reached the volcano's mouth; the others had
been drawing charts of the terrain, and were thus delayed. My
father was sure that if he had been there, he would have been
among those who had gone on ahead. It would seem that his
cautions against youthful enthusiasm probably stemmed from
this episode, which he probably related to Kuwabara.

My father spent most of his time on study and research from
the day he entered the Geology Department at Tokyo University
in 1893, until he retired from the Geological Research Institute.
He traveled all over Japan in the course of his studies. He
married my mother, Koyuki Ogawa, in the spring of 1894.
However, he was actually traveling about, searching for crystal
formations, until the night before the wedding.

In one of his writings, he says: "I write about this because I
cannot leave out this part of my life. But our marriage has no
romantic stories, because the first time we met, one was a baby
and the other a child of seven or eight." As the Ogawa family
was from Kishu, my parents were childhood friends, in a sense.
Though they had lost contact after growing up, they probably

shared some childhood memories. When my father was adopted, they probably felt like brother and sister. Even after the marriage, my father continued to live at Hongō.

My mother might really have become a modern woman, even though she seemed old-fashioned and was intent upon her housework, especially as the number of children increased. Her father, Komakitsu, enrolled her in the Toyōéiwa School for Girls. She was one of the few girls of her generation who studied English. After two years she had to leave school, probably because her father had agreed to pay for my father's education. As I look back on it now, my mother had an extremely logical mind and she was not at all superstitious. That may have been the result of the new educational methods adopted by Komakitsu. For as far back as I can remember, there was a copy of the women's magazine, *Fujin no Tomo*, on my mother's desk, and she admired its editor, Motoko Hani. Once I was invited by the editor to speak at *Jiyugaken,* a school she was connected with, and I took my mother with me. She was very happy to meet Motoko Hani.

Only a few days before he was to graduate from Tokyo University, my father received the news that his father was fatally ill in Wakayama. That was July, in the twenty-ninth year of Meiji. My father left for home immediately; when he arrived, his father was still alive, and he wished he could show him his diploma. Standing beside his father, who was still conscious, he tried to hide his tears as he wiped them from his face, together with the perspiration from the summer's heat, and he said, "I have returned. The graduation ceremony will be very soon." Behind Takuji, his mother covered her face with her hands and muffled her sobs.

The concern with the diploma was understandable. There were not many universities, as there are today. The value of a university education was different; the mood of the students was different. Japan, as a country, was increasing in prestige; it was right after the Sino-Japanese War. For a student in that period, graduation from Tokyo Imperial University was the greatest of honors.

After the funeral, my father went to Ayabé in Tanba and participated in a lecturing contest. He returned to Tokyo in August, and entered the Graduate School as soon as he received the diploma. In January of the next year he decided to work at the Geological Research Center; it was for him a new beginning.

His first assignment was to collect volcanic rocks from the Bōsō peninsula. A Dr. Kochibe was to take these samples to the International Geological Conference at St. Petersburg. His next assignment was a geological survey in Shikoku that was to take about four months. But the survey could not be completed in the time allotted, and as there was some money left over, he made his own decision to continue, and stayed another month. On his return to the Research Center, he was reprimanded by the Department Head, Dr. Nakajima. However, the latter was not unsympathetic, and he paid my father's expenses for the extra days.

But Shikoku was not a good place for my father. He was thrown out once when a rickshaw overturned on the way to Kōchi. Three years earlier, when he visited the Awaji Island there, he was detained by the military police. He had come under suspicion, as he had been drawing charts near the Yura military base; that was the year the Sino-Japanese War started.

The thirtieth year of Meiji passed, and it was decided that an International Exposition and Geological Conference was to be held in Paris two years later. My father's proposal that a geological cross-sectional chart of central Japan should be sent there was accepted and he was assigned to organize the data for the maps. Unexpectedly, when he finished the survey, he was told that he was to be a member of the Japanese delegation to the Paris Conference.

How happy he must have been to hear this news! His

selection was due, in part, to the fact that my father knew some French, and none of the others did. Thus, his job became entirely one of preparing for this trip abroad. The delegation left Japan on a French ship on March 3, 1901, the day of the Festival of Peaches; Yokohama Harbor was foggy. The group of ten consisted of scientists and some people from the management office of the International Exposition. Many friends came to see them off. Komakitsu was there, and my mother, Koyuki, followed with her eyes as my father moved about the deck. For my father, it was a triumphant event: at thirty-one, he was the youngest member of the delegation.

It is amusing that my father's first job in Paris was the purchase of silk hats and tailcoats. He was excited by what he learned from the many scientists of foreign countries and he learned a lot of new geological information at the Conference, but his trip was not entirely concerned with geology. He was the sort of man who would recite Byron's "The Pilgrimage of Childe Harold" to the sea while on board the ship returning to Japan. He thought of Florence on the Arno as being like quiet Kyoto on the Kamo River. "The Shakespeare play that I saw in London was not very interesting, but it was the fault of the mediocre actors who performed it," he said later.

The European journey was a great success for my father. The condition for the trip was that the expenses be limited to 2000 yen; no time limit was set. That is why my father found it possible not to return to Japan until May of the following year. He remained abroad for fifteen months; and he resumed his geological survey in Japan only one month after his return. During this period of his life, my father was seldom at home. My mother kept to the house while he was away, but perhaps that was normal for a wife in that period.

In 1902, my father and Unzon Wada (who is sometimes called the father of the Japanese steel industry) began planning an expedition to China. There were absolutely no data available on the mineral resources of China. As their plans were developed, the Ministry of Foreign Affairs decided to back them, and they went to visit the Minister, Ikataro Komura. In May, the survey team, now six in number, headed for Amatsu from Nagasaki.

These accounts show that my father was a much more active person than I am. In the last years, I have traveled often, but usually because I have to. A man must often move about because of the circumstances in which he finds himself. So, occasionally, I fly to America or Europe, and I visit Tokyo or some other city once a month. But my personality tries to avoid trips. Perhaps I am lazy; I am happier thinking in my office or my study than being on a trip. Theoretical physics, simply stated, is a science that tries to find the teachings hidden at the root of the universe; it is close to philosophy. Geology, on the other hand, is very close to natural phenomena. Our choice of careers shows the difference in temperament between father and son.

My father's trip to China took about a year, and it took place in the midst of international tension that resulted in the Russo-Japanese War. When he returned, my mother had already given birth to my oldest brother, Yoshiki. It was his third child, and the first boy. It seemed that my father's way of life would slowly change. At about this time he began to study *go* with the go-master, Kei Iwasa. "That was when I learned to smoke," my father said. "It was awkward for me in China, because I could not smoke." Perhaps he was not so hopeful about his *go* from the beginning, but I am not sure. He certainly bought a great number of books on the subject. He never became a good player, but he did keep up the habit of smoking.

My father's very busy life was not yet over. The Russo-Japanese War finally broke out in February 1904, and victory was continuous from the very first day. The Endai Coal Mines fell into Japanese hands in September, and it was necessary to test the geological condition of the mines. My father had to cross the sea again, this time by military order. That was another major event in his life. He was on the battlefield with the soldiers. His duty was geological, but it was not like peacetime. He never talked about those experiences, and they could not have been easy.

Soon after he returned from the war, he was appointed professor at Kyoto Imperial University. After ten years of active service in the Geological Research Institute, a quiet life of lecturing awaited him. He traveled now between the classroom and his home in Kyoto. At that time, there were brilliant people at the University, like Konan Naito and Ikutaro Nishida. My father was ill several times, and he would pile up his books beside the bed and read happily. I remember his expression at those times. He often told us about Europe, and was especially proud of the medal he was awarded in France. I cannot forget his smile, as he showed us the medal.

Chapter 3

Iwan (I won't say)

Very old and very new things co-existed in my house during my childhood.

My maternal grandfather, Komakitsu, was a samurai who spent all his days at Wakayama Castle, before the Meiji Restoration (1868). His learning in Chinese classics was enormous, but he studied European culture after Meiji and habitually read *The London Times.*

My father was a modern scientist who went to China and Europe for visits and research on various occasions. At the same time, he was taught Chinese culture from his early years, and felt a closeness towards Chinese literature. He loved old things: old books, antique pottery, and stone Buddhas. He was deeply interested in archaeology. Often he was away; but when surrounded by his family during dinner he would talk cheerfully to us. Once my father looked at the children and said, "You will have to go to Europe when you grow up." I do not recall what my brothers replied; as for me, I had no desire to go to foreign countries.

I never felt a yearning for foreign countries, but I never spoke my feelings openly. I could not speak in front of my father because I was afraid of him. Today, I continue to try to avoid trips overseas unless absolutely necessary. I am lazy, and often I think human relationships are tiresome, even among Japanese. Relating to foreigners simply wears out my nerves.

My grandfather would go for a walk within Kyoto almost every day. He used to go to the market place *Nishiki* and buy his favorite foods for the dinner table; sometimes I went along with him. The downtown streets were covered by demountable cloth canopies to shade them from the sun, forerunner of the present arcades. Little wind-up drummers played tiny drums ceaselessly in front of a toy shop with a small water fountain. The movie house called Imperial House played samurai films and was very popular with the children.

My own favorite pastime was making miniature gardens. I would take a flat rectangular container and fill it with sand and moss. Then I would place bridges, farmhouses, and shrines until the small world was complete. I enjoyed that.

One day — I was perhaps five or six — my father asked my grandfather to start me on the *sodoku*[1] of the Chinese classics. I abandoned my dream world that day and stepped into the world of two thousand and some hundred years ago, left on the pages written in indecipherable *kanji.*[2] We speak of the Nine Chinese Classics. Its first part, the Four Books, starts with *Great Learning,* and that was the book I started with.

The *Analects* of *Confucius* and *Mencius* were at the beginning also; but all of these books were like walls without doors. Each *kanji* held a secret world of its own; many *kanji* made a line and several lines made a page. Then that page became a frightening wall to me as a boy. It was like an enormous mountain that one had to climb. "Open Sesame!" would do nothing. I had to face this every evening for thirty minutes to one hour.

[1]Practice of reading without the necessity of understanding.
[2]Chinese characters.

Grandfather sat across the table and pointed at the page with his *"kanji*-pointing" stick. He read as he pointed to the characters and I followed him, "Shi, notamawaku ..." It was a complete *soduku*. I was afraid of it, and even afraid of the stick in my grandfather's hand. It was like crawling in the dark — what touched my hands was unfamiliar. I was tense, and the tension brought on fatigue. Suddenly I would feel sleepy and go into a strange trance. Then my grandfather's stick would strike a character on the page sharply, and I would hastily concentrate on the reading. It was a trying experience and I wanted to escape.

My feet would fall asleep on cold nights, and on hot evenings, sweat ran down my back and made me uncomfortable. But sometimes my fancy would leave the book before my eyes and take flight, and I would follow my grandfather's voice only mechanically. One night I became aware of the sound of the rain striking the eaves, as I sat before my grandfather, and I suddenly thought of the small spiders. There were many trees around the small shrine in the back garden, with nests of the samurai spiders at the roots. The nests are like small tubes that go into the ground. If you, ever so carefully, pull up that delicate nest, you can find the little spider hanging on to the end. These spiders are called "samurai" because sometimes they kill themselves by cutting their stomachs open when they are caught by humans.

Why did I think about those spiders while reading *Great Learning*? The spiders lost their hiding places when someone pulled up their nests. Perhaps I thought I was like those spiders; or perhaps I longed to escape into the moving world of spiders from the world of immovable Chinese characters.

The sound of the rain continues — I wonder what is happening to the spiders; but *sodoku* is not about to end. The stick in my grandfather's hand follows the characters accurately. I peek at my grandfather's hand that holds the stick; it is old and withered. His beard is long and white and seems to glisten.

It is probably the first time I think about old age. Grandfather is sitting rigidly; he is kind, but not the type to leave duties undone. So he continues until the time is up — no, until the planned lesson is finished — without changing his expression.

I do not think my time was wasted by these readings. The *Toyo-kanji*[3] instituted in post-war Japan are effective and necessary to reduce the work load on children's minds; the labor of memorizing *kanji,* directed elsewhere, can do good. But in my own case, the Chinese classics that I read without understanding have been a great gain; because I had become used to *kanji* through the classics, I had no difficulty in starting to read different materials.

I cannot recall what I read before starting the classics. It must have been picture books and comic books, but I have no clear memories about them. The only exception is the *Kodomo no tomo,* a magazine for children. My mother, who read *Fiujin no tomo,* gave her children the magazine of the same editor, Motoko Hani. Copies of the magazine were always on Mother's table and the children took turns reading them. The magazine was largely didactic, as I look back now, but it was different from the moral atmosphere of the time in that it tried to give the disciplines necessary within the society, as well as in the home. The labor on the part of the editors must have been enormous. The boys in the story were always named Uetaro, Nakataro, and Shitataro (i.e., up-son, middle-son, down-son). The girls were named Koko, Otsuko, and Heiko (like: Anne, Betty, Cathy). These names are enough to make me shudder, but in retrospect, it was extraordinary the way these stories taught valuable patterns of behavior, and made it interesting at the same time.

Sometimes I asked my mother about these stories: What do they mean? Mother would always stop what she was doing. (That is an impressive attitude in a parent. She never said she

[3]Simplified Chinese characters.

would tell me later, but would look straight at me and give an accurate answer right away.) How beautiful her eyes looked at times like that! My brothers and I were much influenced by the stories, and it was my mother who planted the capacity in us to enjoy them.

My mother planned academic careers for all her sons — I was aware of this even as a child. Probably, without my mother's efforts, so many of our family members would not have entered academic fields.

Even as a child I was fond of solitude. I felt a deep antagonism toward my father, and was also afraid of him. Those feelings closed my mind, but my imagination took flight within my closed world. The books that filled our house gradually began to capture me and to feed my imagination. There were books on many subjects, as my father had many interests; especially, there were literary works.

The books I loved best were the ten volume edition of *The Chronicles of Taiko*.[4] I read attentively these beautifully illustrated books in old-fashioned bindings. Because of my grandfather's tutoring, I could read them. I believe I had entered the elementary school by the time I finished reading these books, which took a considerable amount of time. As a result of their influence, I came to love Toyotomi Hideyoshi, drawn by the character of this man, who had the greatness of an explorer.

Still I was shy, and many people remember me as a quiet and timid boy. I can remember that I never spoke much; all complicated matters I handled with, "Iwan" (I won't say). I hated to explain, and became silent if someone asked why I had done something. At times the silence itself was an act of insubordination to my father. So I was not basically timid; "Iwan" perhaps hid my resignation. I even began to feel antagonistic toward

[4]Taiko was Toyotomi Hideyoshi, the imperial advisor (i.e., virtual ruler) of Japan in the 15th century.

Confucianism. My attempt to protect my reserve brought me the nickname "Ian-chan."[5]

Around this time — near the end of the Meiji era and the beginning of the Taisho era — Japan was permeated with Tolstoy. The Shirakaba Group suddenly arose; works of Tolstoy were translated one after the other. At one point, there was a monthly magazine called *Study of Tolstoy*. Is there another example of a foreign author who became the subject of a monthly magazine? The adoration of the Russian novelist was greater than can be imagined today. The "New Village" was built. Sumako Matsui's "Katusha" became very popular, and her song was on the lips of every girl. Perhaps (I think so), the nickname "Ian-chan" was taken from "Ivan the fool". If so, it is a humorous, but dishonourable, name.

What I admired most about my mother was her effort to be fair to all the children in our large family. My younger brother was physically weak and was carefully looked after. He often had a high fever from flu and from ear infections. When he was ill, special food was prepared for him. I envied the special treatment my brother received and wanted to taste the eggs he was eating in bed. One day I also caught flu and my mother cooked eggs for me, but they did not taste particularly good, as I had lost my appetite.

My mother began discussing things with me about the time I began to attend high school, perhaps because my next older brother had the tendency to ignore her. She did not speak much at all, and I often found her deep in thought. It seems to me that she was exceptionally rich in contemplative faculties, for a woman. She was never fussy towards her children, although our upbringing was basically strict. We never spoke emotionally to my mother, and my mother never tried to explain her needs.

[5]"chan" often follows a name familiarly, especially among children.

Most of my clothes were hand-me-downs from my brothers, which cannot be avoided in a family with so many children. But I am convinced that Mother tried to be fair to her children in such things as food. I will write about the sweets. The children received sweets at three o'clock every day. There was a candy shop called "Kakiya" at the corner of Teramachi and Nijo. For that period, the store was modern both in merchandise and architecture. Almost every day, the store sent someone to take orders. The man wore a striped kimono with a striped apron and brought wooden boxes with samples of sweets inside. Mother gives the order; the man writes it down in a notebook, wraps the boxes up in cloth; the man writes it down in a notebook, wraps the boxes up in cloth, and leaves saying, "Thank you". The children were excited with anticipation, although we rarely had expensive sweets. Salted and sweet beans, and *suhama* — not those sweets sought after by gourmets. We were given equal portions of these ordinary sweets and we looked forward to it.

There were cotton candies and many other delicious-looking things at the festivals of the Nashiki Shrine. I always wanted to buy them, but knew I was not permitted. The cotton candy that flies from the middle of the copper container like white thread always awakens nostalgia in me.

There is something I have noticed: that a man's attitude changes many times with age and environment. I have written about my parents with candor, but my childhood memories of my parents differ considerably from those of my sister Kayoko. My father seems to have helped my sister with her school homework; for example, he looked in the encyclopedia for material on Harriet Beecher Stowe. When my sister had to write the genealogical table of the family, it was my father and my grandfather who became preoccupied with it. The old records of the family were taken out of the Chinese leather boxes and scrutinized. My sister was very sleepy for several nights because she had to write a very long report. They took her to many exhibitions and told her their criticisms, much too difficult for a child's understanding. According to my sister, "One of the reasons that we moved to Kyoto was because my father wanted to spend more time at home for the education of the children."

At the same time, Mother also possessed a side that I never knew. Her youth was spent in a new age, and she was educated in a new way by Grandfather Komakitsu. At that time, European thoughts spread through Japan with amazing speed. Voices calling for the liberation of women began to be heard, and occidental clothing for women began to be popular. In the photographs, Mother was a pretty girl with beautiful eyes. She was wearing a dress; she was probably an active girl, who was permitted such freedom by my grandfather.

I am told that my mother liked to ride on the swing. Perhaps, as she sat swinging, she enjoyed the sense of escape from the traditional values that had made prisoners of Japanese women. In the old-fashioned days, this might have even been considered a revolutionary gesture. How beautiful and free a girl in a dress on a swing must have looked to those people! But she made one mistake: once, when she was still young, her hand slipped and she fell from the swing, striking her head. She crouched on the ground and was unable to move for a few minutes. The accident seems to have shocked her, and later, when she was plagued by headaches, she recalled the fall from the swing.

On her deathbed, in the eighteenth year of Showa, Mother asked me to arrange for an autopsy of her brain after her death. That was when she told me about the swing accident. Although she was the wife of a scientist, and had children who were scientists, I was moved by her having such a calm state of mind before her death. The autopsy was performed by Dr. Teizo Ogawa, who is a brain specialist and also my eldest brother's close friend. It was reported that there was no injury and that the brain was considerably heavier than the average.

Mother stopped going to the Toyō-éiwa School for Girls after my father joined the Ogawa family. Dropping out of school was not unusual for girls at that period. After school she learned the usual women's studies of *koto,*[6] Japanese songs, flower arrangement, and tea ceremony. She also attended lectures on *The Tale of Genji.* The lecturer said that *The Tale of Genji* is like *miso* soup,[7] according to my sister. She thinks the instructor probably meant that *Genji* has a particular quality that one does not tire of, even if one reads a small portion each morning.

Mother seems to have remained active even after marriage, for as long as the family lived in Tokyo. Father was often away on trips; there were not yet many children, and Grandmother was able to look after them. Mother took cooking lessons every week; these lessons received hearty support from the children, who waited eagerly for the boxes packed with the delicacies she brought home. Sometimes Mother took my sister to events held by the Patriotic Wives' Association.

During those years, Mother participated actively in the children's education, giving evidence of a young and intelligent mind. The English textbooks Mother had used at school had titles like *Kanda Reader* or *National Reader.* She kept these books and later taught Kayoko with them. Kayoko was then in the higher levels of the Azabu Elementary School, which already had English as a mandatory subject.

I am told that Mother sometimes showed literary magazines of the late Meiji period to my sister, who says, "My interest in literature seems to be in large part a result of my mother's influence." Mother's magazines were thin by today's standards, but contained literary works rich in content. My sister became familiar with names like Bimyosai Yamada (a Meiji novelist) through this magazine. I had always thought that my interest in literature was developed through the many books that filled the house, but after hearing my sister I wonder if, in my case also, my mother was not in a large part responsible.

---

[6] Japanese musical string instrument.
[7] Bean-paste soup, served at breakfast.

Why did my mother's behavior change so completely after we moved to Kyoto? She did not dislike leaving the house, but rarely went out except to a school play or a school athletic competition. Part of the reason must have been that her hands were full, with so many children, but the character of the town of Kyoto might have had something to do with it. It was a Kyoto custom for the wife to remain in the home, a custom that probably still runs in the blood of the people. Even today a wife tends to decline an invitation, even if both she and her husband are invited. Mother's change seems to have taken place after we moved to Kyoto, and it was under her influence after her change that I grew up.

Chapter 4

Somedono

Meiji — the name reminds me of water in a flask over an alcohol flame, gradually heating up and boiling over. My early years were spent toward the end of that period, but I cannot recall how the Japanese people mourned the passing era; I was only a little more than five years old. My older brothers probably knew many stories of Emperor Meiji; the only thing in my memory is a book that my mother gave me. It was a fold-out picture of a long funeral procession of people in full court regalia. I looked at it every day without comprehending its meaning; I did not know that The Meiji Era had ended and the Taisho Era had begun.[1]

In April of the second year of Taisho, I began elementary school. Although our house was now in a different school district, I attended the school that my older brothers went to. It was considered one of the better schools, with a high proportion of students planning for higher education. Thus, many professors sent their children to the Kyogoku School, and not to the other, the Kasuga School. The students of Kyogoku used to taunt the students of Kasuga, by stressing the "kasu" — meaning "garbage". But as I never heard of any fight resulting

[1]Meiji, 1868–1912; Taisho, 1912–1926; Showa, 1926-present

from these taunts, it may be that Kasuga really had the better students. On the other hand, the Kyogoku School students called their own school the "raining school". Whenever the weather was good and the teachers decided to have a school picnic, it rained on the day of the picnic.

Kyogoku School is located in Somedono. Strictly speaking, it is not in the Somedono district, but its school song, composed after my graduation, speaks of Somedono. The school is not far from the house we lived in before we moved to Kawaramachi Street. The school building is of concrete now, but in those days it was of wood. The oldest elementary school in Kyoto was opened in May of 1869; sixty-four other schools were also opened that year, when there was an explosive growth of the educational system. Kyogoku was one of those schools, it was then called Kamikyo 28th and 29th Combined School. It changed its name when it moved to Somedono, and after its reconstruction, in the sixteenth year of Meiji, it became, finally, Kyogoku.

All of Kyoto is full of historical landmarks, but the section just to the west of the Imperial Palace is particularly rich in relics of the Heian Period.[2] Somedono takes its name from Somedono-in, the site of the palace of the Fujiwara clan. There is also a place where Lady Murasaki (who wrote *The Tale of Genji*) is supposed to have lived.

Elementary school was my first contact with the larger and more complex world outside my home, and I cannot say that I made a good start. Probably I inherited from my mother a tendency to be silent. More than other children, I was given to anxieties, but my worries were not about real dangers lurking in the world outside; rather, I was a meek lamb, peering through a half-open window. Sometimes I think that I must have had a lower psychological age than my actual age. Once, however, a girl in my class and I were selected to take a psychological test

[2]*Heian* was the name of Kyoto at the time of the domination by the Fujiwara family (794–1185 A.D.).

from Professor Katsujiro Iwai of Kyoto University. I remember walking home through the lonely campus at twilight after taking the test. I felt like a guinea pig who had been used in a strange kind of experiment; later I was told that I had a high IQ.

My two older brothers were in the sixth and the third grades, and I used to walk to school with the older one. From our house in Kawaramachi, we turned left, walked toward the gate of the Imperial Palace, then turned north on Teramachi — a walk of perhaps ten minutes. A street car line ran on the west side of the already narrow Teramachi Street, passing very close to the walls of the houses. The students always walked on the east side of the street, with the teachers taking turns in directing traffic in front of the school to protect the arriving children. But thinking back, the train was not the sort likely to injure people. It was small and, above all, it was slow. The conductor could jump off the moving train, chase people off the tracks, and jump back onto the train. The same was true of the horse-drawn carriage of the mayor of Kyoto. When pedestrians blocked its way, the footman would jump off the carriage and run in front of it, clearing away the people. When the path was cleared, he jumped back on, agilely and elegantly. Those were the days!

Even the gate of the school presented a certain dignity to a new student. It was a gabled gate, designed and built five years earlier, after one in Momoyama Castle. It could even cause fear in a sensitive six-year-old boy. The gate was sold later, I am told, to a factory, and now stands beside the freeway near Tōji.

My class was the "I" class of the first grade, which also had a "Ro" class, about eighty students in all.[3] First and second grade classes had boys and girls together. Behind the main building, which stood close to the gate, was the exercise yard. To its left was a flower garden, and behind that were the classrooms for science, home economics, and music. To their left was a pond in the shape of the island of Honshu, and behind the pond was a

---

[3] *I* and *Ro* are the first letters of the Japanese alphabet.

cherry tree. Every morning there was a ceremony, in which the principal stood beneath the cherry tree and made his speech. Our *alma mater* had not yet been written, so we sang, "The Diamond":

Diamonds must be polished
Or the gems will not shine.
Only after learning
Can men know their true virtues.

It is a dear song of the Meiji Era — no, perhaps I am going too far here. I have no nostalgia for Meiji; I was born in its fortieth year.

Across the street to the west of the school was the Imperial Palace, whose foliage I remember as greener and more beautiful than at present. My classroom was in the middle of the north wing of the building; the cherry tree could be seen nearby through the window. Boys and girls sat at alternate desks, and next to me sat Mikiko Narikawa. She was slim and pretty, and quite small. We quickly became friends, sitting together. Aside from relatives, she was the first girl with whom I became friendly.

I remember the clothing of the students: girls wore *kasuri* (splash-pattern) kimono with *heko-belt,* while some wore *hakama* (wide pleated trousers). Unless there was a special occasion or ceremony, most of the girls wore pinafores over their clothing. A few from rich families, I suppose, wore overcoats (*hifu*) during winter. In my mind, I can still see the violet hue of the decorative ribbons on the kimono. Most of the girls had their hair in braids, some also decorated with ribbons. Most boys wore blue *kasuri* kimono and sandals with leather soles. At one time, it was the fad to put metal rivets in the soles, but the school forbade them, because of the noise they made.

I wore *hakama*; not many of the boys did. I carried my sandals in a bag, because on the way to and from school, I wore *geta* (wooden clogs) on my feet. Sometimes I put the sandals in the same bag with my lunch box — not very sanitary! Everyone was impressed, though, by the boy who arrived at school each morning in a private rickshaw, and who wore Western-styled clothing, which was a very rare thing for children. That boy was the son of a viscount, named Kiyoyasu Higuchi.

Our teacher's name was Kawamura. Our Japanese reader had old-fashioned pictures of objects with their names written above them in *katakana* (one of two Japanese alphabets): flag, kite, top, dove, beans, etc. That was rather boring!

For a child who had been reading Confucian texts every night, first year Japanese reading was simple. While I listened to the teacher read the text, I was distinctly conscious of the bright color of the cherry blossoms on the tree outside the window. Then suddenly I heard, "Ogawa!" I had been asked a question, and I stood up. Somehow, although I had heard the question, I could not bring myself to give the answer. (I had this problem often after that first incident. Perhaps my father's hot temper had implanted the fear that I would be shouted at if I gave a wrong answer. Perhaps my silent tendency, inherited from my mother, was responsible. In any case, it took me many years to overcome the inability to begin to speak, and the tendency is still a part of me.) I knew that every eye in the classroom was on me, and I became red in the face as I looked at the teacher and nervously fingered the folds of my *hakama*. Mikiko, sitting next to me, whispered, "You *know* the answer," but I only became more red-faced.

Outside, during the break, Mikiko was collecting fallen flower petals under the old cherry tree. She was stacking them together, piercing them with a pine needle. As she noticed me approaching her, she offered them to me, "Would you like to have them?" I accepted her offering in silence, and when she asked, "Where do you live?" I managed to say, "Teramachi-Imadegawa." Silence again, then: "My brother and your brother go to the same school." My eyes were drawn to her earlobes; they were just like the cherry blossom petals she had given me.

One day, Mikiko was asked by the teacher to solve a problem in arithmetic. She stood up, but could not answer right off. Hastily, I wrote it on the edge of a notebook page and pushed it toward her along the desk. Blushing a little, she glanced at the number, answered, and sat down. Her eyes were shining as she turned toward me, and beyond the shining eyes I could sense a warm feeling.

Our desks were close enough to allow these interactions. They were arranged in pairs, with a boy and a girl at each pair. The desks were old and dirty, of the type that could be opened by lifting the top in order to place books and notebooks within. There was a small drawer in the right-hand corner to keep the brushes and the ink-wells for calligraphy. My desk was always exceptionally neat. Even at home, I was the kind of child who could not settle down to work at the desk, if it was not placed parallel to the lines of the tatami mats. I had to work for many years to overcome this kind of anxiety.

Mr. Kawamura, my teacher, was still a young man. He wore a black coat and a very high white collar. This was, perhaps, the era's *haikara* (the Japanese version of "high collar", that came to stand for "new", "progressive", European", etc.). On the day when ceremonies were to take place, the male teachers wore frock coats. Many of the gentlemen sported pointed moustaches in this period. A teacher in a frock-coat with a pointed moustache possessed a certain dignity that made him appear unapproachable to a child. The women teacher wore kimono with long sleeves and *hakama* with many pleats. Most wore their hair in high sloping coiffures. That sort of dress made the teachers seem older than they were. Today's teachers in modern clothing look much younger.

All through elementary school, and until I entered high school, I was smaller than my classmates. My skin was dark, I tended toward plumpness, and my appearance was childish and innocent. Something called "psychological observation" of the children was practised in my school. Later, I learned that my teacher in first grade had written of me: "Has a strong ego and is firm of mind." He had seen past the innocent face to the sensitive nerves and the competitive spirit.

I was slow getting used to the school, and made few friends. The children I played with at school, I hesitated to invite to my home. The problem was related to my being unable to speak out. I could not bring myself to say: "Why don't you come to my house?" I did not develop socially (and never would, really) partly because it was an inborn trait, and also because I had several disappointing experiences very early in elementary school.

As soon as I began school I became a close friend of a boy named Endō. I remember that we used to run around the gymnasium together. His father was a policeman, and apparently was assigned to a new post, because Endō had to change schools. That was my first disappointment. The next close friend I had was named Jō Nakamura. He was quiet and very intelligent; his family lived in the Honzenji Temple on Teramachi Street, and I used to go there to play with him. (That was the temple with the "Laughing Emma".) But near the end of first grade, his family also moved away.

During my second year, a girl named Hisako Uchié was transferred into my class. She was both exceptionally smart and especially pretty. The other girls complained that the teacher, Suwa-san, was partial to this girl, but it seemed to me that she deserved such favored treatment because of her cleverness. But she, too, had moved away by the end of the second grade. Probably, I lost interest in making friends because these successive incidents occurred during the first two years of elementary school.

I was so methodical-minded that, at home, I insisted that the desk had to be aligned parallel to the tatami mat stripes. It was not that I studied at that desk; there was no homework, nothing to review or preview. At home I either read whatever I liked or played outside.

It must have been lonely for my younger brothers, Tamaki and Masuki, after I entered school, for they were always waiting for me outside the gate when I came home. But they were not very good playmates for me. They would not play "catch" nor "shotput". They were very quiet children; Tamaki, especially, was a bookworm, and was often to be found looking through books, even before he could read.

This used to happen after Tamaki began elementary school: My father might say: "Bring me such and such a book," addressing nobody in particular. Yet it was invariably Tamaki who got up to get it. It was not only that he was obedient when father called. He *was* obedient, but more than that, he knew the position of every book in the house, whether or not he could understand its contents. He had a brilliant memory and had memorized the title of every book in the house, and all he needed was the title — no need for Father to specify the room or shelf.

As far as interest in books was concerned, however, I was no less enthusiastic than my brother. After I had finished *The Chronicles of Taiko*, I started reading translations of *Anderson's Fairy Tales*, the Brothers Grimm, and the like, and children's stories by Sazanami Iwaya. I loved magazines like, "The World of Boys" and "Japanese Boys". Names of writers I remember are Hosui Arimoto and Shisui Matsuyama, and I enthusiastically read Miekichi Suzuki. The magazine, "Red Bird," that began publication while I was in elementary school, captured my heart.

I preferred a now forgotten song called, "Back and Forth" to the well-known "The Canary that Forgot How to Sing", and sometimes I hummed the tune to myself, without thinking:

Back and forth, yesterday and today,
The white clouds over the mountain tops ...

It was a more sentimental song than "The Canary", and sentimentality is still largely a part of me. I was very fond of the
works of Ruiko Kuroiwa, and I had a beautiful pocket-sized
edition of his translation of *Les Miserables.* This story moved
me in a strange new way that I had not experienced before.

There was no modern Japanese literature among my father's
books, but I did read Kōyō and Soseki. I read right through the
collection of classics of the *Ariakedō-bunko* (a publishing
house). I read almost every day, almost everywhere, so that I
practically memorized the names of numerous literary
characters. It seems I could understand the *Tale of Isé* and the
Tale of Héike, but it was not until I entered the middle school
that I began to enjoy seventeenth-century works like those of
Chikamatsu, Saikaku, and the plays of Joruri-style.

I also read foreign works, Turgenev and Tolstoi, as well as
French and German novels (in translation). But the one foreign
author who commanded my attention above all others was
Dostoievski. In Japanese literature, it was Chikamatsu. (Is there
some link between these two authors?) In any case, I was a boy
deeply interested in literature, who showed no indication that he
would become a physicist.

Through my wide reading, I became even more introverted;
but my eyes were not entirely closed to the outside world. In
front of my house was the Kuninomiya Palace, and to its left
was a hospital. The house next to our own was owned by the
former principal of the Third High School. To the west was the
house of Mr. Toyōko, our landlord, and farther west lived Mr.
Koyama of the Museum, the younger brother of Haktei Ishii.
Beyond that was the former Minister of Commerce, Mr.
Kataoka, and then Viscount Takakura, whose beautiful
daughter was chosen to be one of the dancers at the coronation
of the Emperor.

The neighborhood was then a quiet suburban area, where each inhabitant lived alone, not communicating with the others. It was a good environment from the viewpoint of educating children. But I was still in the lower grades of elementary school, and I was drawn toward the children's world which was a little farther away. There was a Buddhist Temple in Demachi Street that held a carnival twice a month. I still remember the smell of the acetylene lamps that were used to light the stalls of the merchants. Also, two streets south of our house there was a Kojin Shrine (Shinto god of the kitchen) that also held festivals with many stalls.

Not only were there stalls selling trinkets and other merchandise, but also picture shows to attract the children. Peering in through a small window to see the pictures would cost, I believe, two sen. A narrator told the stories that he accompanied with the rhythmic tapping of a stick. I was too small to understand the stories, but I liked the atmosphere of the picture shows. Next to the picture show, there would be a street singer offering popular songs. There were goldfish shops, and also shops that sold ground cherries, Kintaro candies, shadow-pictures; during the summer there was corn-on-the-cob. There must have been stalls selling household items and cheap kimono, but all I remember are the things that might interest a child.

It might have been while I was returning home after attending a festival, that I saw children playing in an alley-way with "bei" tops. The tops were made of iron, perhaps two centimeters in diameter, and they were spun on a mat placed on top of a wooden box or a large bucket. Sometimes the tops touched and sparks flew, and one top would sometimes be knocked off the mat by another. The children went wild with excitement; they exchanged harsh and fiery words, and did not even notice when the sun had set. They realized that night was approaching only when one of the tops became lost in the grass and could not be found because of the darkness.

There was another children's game called *menko,* a *menko* being a disk of cardboard with the face of a well-known actor or military hero painted on one side. One *menko* was placed on the ground and another thrown at it. To win, you had to turn over the one on the ground with the impact of the one thrown. *Menko* soon became scarred and dirty; General Nogi's moustache was shaved off, and General Tōgō had a hole in his forehead. A similar game was played with *kanamen,* small plates of lead shaped like airplanes or dirigibles. One *kanamen* was dropped on another lying on the ground, and the aim was to turn it over. However, I was allowed to have neither *menko* nor *kanamen.*

I was filled with envy when I saw a boy with dozens of *kanamen* tucked in his belt; what I envied was his freedom. Boys who played with these toys were "city-kids"; most of them were the children of merchants. However, I was not entirely capable of understanding why "city-kids" were allowed this freedom, and we were not. The only person who brought me closer to city life was my grandfather, Kamakitsu. And though I suffered during his nightly Confucian lessons, I showed my joy openly when grandfather would say: "Let's go, Hideki." I would ask: "Where, Grandfather?" And he would answer: "How about Shinkyogoku?"

Shinkyogoku was the liveliest part of Kyoto, something like Asakusa in Tokyo, where the Kannon Temple is located. The street was narrow, and it still is. Stores lined the street; several theaters were located there. There were not as many restaurants then as there are today. It was a relaxed and friendly place with a busy downtown atmosphere. I would stare in wonder as I walked, holding on to the sleeve of my tall white-bearded grandfather. There were housewives, and older men who looked like tourists. Women in elaborate Japanese coiffures walked from shop to shop. The warm breath of life saturated that part of town. The sales-calls of the shop attendants were loud and urgent. Was it in front of some shop that I saw the red and green banners flying, or was it at the wooden doors of a theater?

Street music could always be heard.

The child's heart danced, though he could buy neither toys nor food. The boy's strangely depression-prone mind seemed to open up. His imagination grew in that atmosphere, so different from that of his home or school. There was even a small bookstore — perhaps it was really a toy shop or candy store.[4] It sold books about the size of today's paperbacks, with paper of poor quality, but with brightly colored covers to attract the eyes. They belonged to the series, *Tachikawa bunko,* with titles like, *The Ten Heroes of Sanada,* and they were to be read for pure entertainment. They were also sold at the festivals near my house, and though I rarely bought them, I did borrow them from my friends.

Reading the classics and foreign novels, on the one hand, and these entertainment books on the other — that is truly a mixed selection, but perhaps that is how it should be for a child. Because his interests are unrestricted, he is open to everything. He wants to make whatever is around him his own, absorbing all that can be absorbed. The chaotic material gradually becomes ordered and organized into his life's direction. Is one's personality not shaped in this way? As I have written before, there was little evidence in my early life to indicate that I would become a physicist. My motives for choosing a career in theoretical physics (though this is far ahead of the story) were largely influenced by chance events.

But I am straying from the track; for me, Shinkyogoku was the living spirit of *menko,* to which I was strongly attracted. But after I entered middle school, my parents forbade me to go to Shinkyogoku.

It was the custom at Kyogoku School to reorganize the classes at the beginning of the third grade, separating the boys and the girls. My new teacher was a man named Shin Shiojiri. He was

---

[4]A *dagashiya*, that sold many kinds of things, which were generally inexpensive.

about thirty, and was tall and stern. His nickname was *Namba,*
meaning, corncob in the dialect of Western Japan. That was
because of his red moustache; but I never called him by his
nickname. He was my teacher until I graduated, and I seemed to
enjoy his deep confidence.

There was a boy named Heisaburo Kuromoto in my class,
whose father owned a sweets shop. He was rather short and
round, and it was he who lent me the entertaining *Tachikawa
bunko* books. One day he was called to the teacher's quarters.
He was wondering what it was all about when one of the female
teachers, named Yamada, whispered the questions: "How
much are the Koshidaka cakes?" The boy was surprised but he
answered: "Fifty sen for each set." (Koshidaka cake is
sometimes called wedding cake and consists of a red piece
and a white piece, making up a set.) "Miss Yamada is getting
married!" The rumor immediately spread through the class.
Her bridegroom, it turned out, was my teacher, Shiojiri-san.
That sort of marriage, between two fellow employees, was very
uncommon in those days. The boy, Kuromoto, still lives today,
behind the sweet shop that he inherited from his father, and he
still retains his child-like face.

I was once chosen by Shiojiri-san to give a recitation, but I
was not able to pronounce a syllable of it and had to step down
from the stage, blushing. It was part of "not being able to
speak". But I was always head-of-the-class, and wore a badge
with a red ribbon. Mikiko Narikawa, by then in a different
class, also wore this badge of honour.

I was never very good with my hands: art, gym, and industrial
arts were not my favorite subjects. Had I been cleverer with my
hands, I would have done better at experiments, and would
probably not have gone into theoretical physics. I never did well
at athletic meets, but once I did take first place in the obstacle
race. I was not fast, but I negotiated the obstacles smoothly,
perhaps by luck. The only thing I did well with my hands was
calligraphy. When I turned in a required piece of calligraphy, on

the occasion of the New Year, my teacher praised it to the class. "Ogawa is very good at it; he has a good teacher." In fact, I learned calligraphy at home from a tutor named Kyozan Yamamoto, who had studied the subject in China.

Yamamoto-san taught calligraphy to all my brothers and sisters, and I used to go to lessons even before I entered school, as a kind of escort for my sisters. His house was to the west of the Hamaguri Gate of the Imperial Palace. Once each week, I left my house and passed through the Seiwain Gate with my sisters, Kayoko and Taéko. They were much older than I, but it was the custom of the day that men and women never walked together, even if they were brothers and sisters. If my sister walked on the right-hand side of the street, I walked on the left side, and never thought twice about it.

Later on, Yamamoto-san used to come to our house, probably because the number of his students increased. He was a portly man with a round face, who had gone to China when he was young to study calligraphy under Shukei Yo, one of the Hokuhi branch, through whom an important tradition of calligraphy had descended. Yamamoto-san taught his students in this way: he would hold the end of the student's brush from across the desk and would write the characters upside down. The student would be holding on to the brush, also, and thus he learned the teacher's movements.

In this manner we would write each character on a single piece of paper. We used these as examples to copy, and after practising, we would make one final copy of each character to turn in to the teacher at the next lesson. My brothers and I never used to practise until the day of the lesson. On several occasions we made *only* the one final copy, and that one while Yamamoto-san had already arrived at the house and was teaching our sisters. To avoid being betrayed by the wetness of the ink, we dried the pages before a fire, which scorched the ink, turning it brown. Because we had no other copy, we had no choice but to turn them in. Yamamoto-san looked at them but only smiled. He used to seat himself regally in his *hakama,* after bowing to us as we entered the room. My brothers appeared to be disconcerted by his manner. Sometime later, this practice of bowing was abandoned. I cannot say I was not flattered when Yamamoto-san said: "You are the best." And, in fact, I had a peculiar stubborness that did not allow me to give up easily once I had started on something.

In later life I learned from my father-in-law to paint *Nanga,*[5] and I learned to compose the traditional poems to go with them. I have Yamamoto-san to thank for giving me the training and the courage to put the latter down in calligraphy. He taught me various techniques: standard, semi-cursive, and cursive. And these lessons continued until I began to attend high school.

Where, then, was the potential that made me pursue the natural science discipline? Mathematics was my favorite subject from elementary school on. I retain one memory concerning this subject. I had figured out, on my own, a method for obtaining the sum of an arithmetic progression, not realizing that this was a slightly higher level of mathematics for an elementary school student. When my brother learned this in middle school, I already knew it. I remember my mother looking at me curiously when she realized this, and suddenly she began to smile in delight.

[5] Southern School of Chinese painting

Chapter 5

A Voyage

The formula for the sum of an arithmetic progression is known to everyone who has learned algebra in middle school, but I devised it by myself. The equation itself was unknown to me, but I was doing the same operation as it expressed. In later years, when my brothers gathered at the house at Tōnodan, they talked with my mother about my boyhood days, and this story about the arithmetic progression came up. My older brother wanted to interpret the discovery as a flash of creativity on my part, but I never thought of it that way; I had almost forgotten that it had occurred. It just happened that mathematics, or arithmetic in elementary school, was my best subject.

For most of my acquaintances, I was not a conspicuous presence from the time I entered elementary school until I went to high school. During the later elementary school years, my grades gradually improved. I received 10's in all my major subjects, and 9's in gym and industrial arts. At the end of sixth grade, I read the farewell speech to the graduating class. The representative of the graduating class was Shu Ogawa, who was to become the wife of my oldest brother, Yoshiki. (Although her name was Ogawa, she was not related to our family.)

In the spring of the eighth year of Taisho, I graduated from

Kyogoku School, and entered the First Middle School. This school later moved to Shimokamo and its name was changed, after the War, to Rakuhoku High School. In those days, it was located south of Kyoto University on Konoé in the Yoshida neighborhood. The Third High School was to the north of the First Middle School.

One pattern for a student in Kyoto was to go from the First Middle School to the Third High School, and finally to Kyoto University. That meant going to the three schools located in the Yoshida neighborhood for ten years. There was no such thing, then, as the "entrance examination hell," and I do not remember working hard to get into any of these schools. Like my brothers, I felt that moving up in this pattern was almost inevitable.

The principal at the First Middle School, Sotosaburo Mori, has been written about by a number of people. Takeo Kubawara, who is older than I, wrote about him under the title, *The Good Educators of the Good Era*. For me, Mori-san was important in many ways. He was the principal of the First Middle School when I first entered it, and later he became the principal of the Third High School, just as I entered it. My Middle School years were shorter by one than those of my friends, but I had the good fortune to learn under Mori-san's direction for seven years. This was very valuable to me, and the spirit of freedom I acquired this way will probably never leave me.

To reach the First Middle School from my house on Kawaramachi Street, one had to cross the Kojin Bridge over the Kamo River. As one approached this old bridge, one could see Mt. Ei directly in front; to its left and lower down was Mt. Daimonji, and still lower was Mt. Yoshida, the mountain beloved of the students of the Third High School, who basked in the joys and suffered the pains of youth. To the right of Mt. Yoshida rose the ridges of the Eastern Mountains. Imadegawa Bridge and the Northern Mountains could be seen by looking up-river, and Marutamachi Bridge could be seen down-stream.

The view around this area has not changed very much to this day. Is it because the mountains and the river occupy most of the field of vision? No, even the houses along the river seem unchanged. There is an old red-brick building in front of the bridge, a textile factory built in the early days of Meiji. The machines in the factory were imported from France, and their use had to be learned in order to start up the industry. Today, it is an old-fashioned building that awakens nostalgia.

As one walked along the north side of the factory, the buildings of the First Middle School came into view, its wall lined by large willow trees. The wooden buildings were old and rickety. One building was supported by a pole propped against it. Siding panels were cracked; roof-tiles were falling off. But the old buildings were also a source of pride at the First Middle School, as it was one of the earliest schools, built in the third year of Meiji (1870).

Principal Mori was kindest toward the students. As I looked upon his gentle face for the first time at the entrance ceremony and heard his unpretentious short speech, meaning: "From today on you are a student of this school, so study diligently," I came to admire him from the bottom of my heart. Although a moderate gentleman, Mori-san had a stern, samurai-like side, as well. I heard much later that certain Representatives in the Kyoto government wanted their children to study in the First Middle School, but they could not pass the entrance examination. It was a more honest age, without the possibility of "back-door" entrance. The Representatives tried to pressurise Mori, and when they realized that he would not give in, they said: "Well, as long as that principal is present, we shall not budget anything for a new building there."

That was the sort of atmosphere that prevailed in the House of Representatives, and that is why rebuilding never took place. When they found out the reason, it only renewed the students' pride in the rickety buildings. It was a sense of justice, unique perhaps to youth. As a middle school student, I did not know the underlying reason, but simply felt pride in the old buildings that represented the long tradition of the school.

I was in the "B class" of the first year. Many of my classmates were children of scientists; many of them became scientists themselves, and occupied positions at the Universities of Osaka, Kyoto, Nagoya, Osakashi, and many others. Sin-itiro Tomonaga, who would walk the same path with me, was one grade ahead of me. My oldest brother was already in the Third High School, but my second brother was in the third year of middle school. My two younger brothers were in elementary school.

When I entered the First Middle School, I found myself surrounded by good friends and good teachers. Many of the older teachers were very learned; many of the younger teachers had graduated from Kyoto University and were destined to leave their marks in the academic world. Some of my fellow schoolmates became artists; as for those who became scientists, there are too many to list their names. Quietly supervising these outstanding teachers and students was Sotosaburo Mori.

Liberalism of the principal, his policy of non-interference, did much to advance the students' education. This was the kind of environment to set free the feelings of an introverted boy. The classmates were interesting; the teachers had a sense of humor. Umakichi Takenaka taught mathematics and was very good at making the students laugh. He was from Tosa and was a graduate in physics. He would bring his slight body to the podium, and then speak suddenly: "Those who are absent, please raise your hands." The class roars at this. When he begins the lecture, he draws a large circle on the blackboard with his chalk — it is a beautiful circle with almost no distortion. He turns to the students and mumbles: "I am *so* bad at drawing circles; I find it very difficult." Of course, he means the opposite; he smiles, and the students laugh. The students loved the mathematics teacher and called him Uma-san.[1]

[1] "Uma" is short for Umakichi, and means "horse"

All the teachers had nicknames, most of them not very obvious in meaning to the outsider. For instance, the art teacher with a moustache was called "Tampaya" because he looked exactly like the owner of a bookstore that had that name. Students who are good at inventing nicknames are probably intelligent ones; they are playful. However, I was never endowed with playfulness, and calling teachers or other students by nicknames always seemed to me to be in bad taste. In the United States, after a short acquaintance, one uses first names, and I despise this custom also. I dislike addressing anyone or being addressed without the use of titles. One of the reasons I admired Mori-san was because he said: "I shall treat you middle school students as gentlemen, and address you using, 'mister'." Although I think the German, "Herr Professor," is too formal and carries an ironic ring, still I have always disliked attempts to become familiar by the use of rude expressions. Perhaps that is one of the reasons I became increasingly solitary.

I would not say that there were no rebellious students at the First Middle School (of course, they were not delinquents). In those days, the students wore gaiters made of white cloth, designed to be held by a button at the calf. The end of the gaiter, if properly worn, made contact with the shoes. Some impudent students wore them deliberately short, so that the color of the socks became visible between the shoe and the gaiter. Some students especially ordered pairs of gaiters that were short, with this end in mind. That was, perhaps, the only form of dressing-up possible for a student of that period.

There were none of today's coffee shops; there were, of course, no students who drank saké. A small group of students frequented so-called "milk-halls"; that is now an obsolete name, but then there were several in the vicinity of the school. The milk-hall was a shop, with glass doors covered with white curtains, and it had a modern atmosphere for that period. Beverages served included milk, milkshakes, cider, and coffee, the latter costing five yen.

There were some disorderly students who, in winter, would sometimes rip off the half-broken sidings of a building and feed them generously into the stove. Broken chairs and desks also became fuel for heating. There was even a worse case: a student stole the chemistry teacher's attendance book and burned it under the floor of the building. For that offence, he was severely punished.

As for myself, I became increasingly quieter as I went through middle school. It is not that I lacked friends; I even participated in various sports. It was just that a boy's inner world had opened up within me; thinking back, I feel that I narrow-mindedly tried to protect myself. The place I frequented most was the library, called *Seishikan,*[2] a building that looked miserable from the outside, like all the other school buildings. Above its entrance was the picture of Konan Naito, who had named the library.

I had a personality that was easily upset. The psychological observation chart at sixth grade contained the comment (along with those like "trustworthy" and "sharp of mind"): "tries not to cry at slight cause, but is quite apt to cry." A heart that is impressionable is easily hurt. To try not to lose my peace of mind, I felt I had to minimize my contact with other people. It was as though I tried to protect my freedom, having as few comrades as possible as I voyaged out on a deep wide ocean. Where I was to set my course, I had no idea. I was a boat without a rudder.

As I said earlier, I was not a conspicuous presence. My friends called me, *Gonbei.*[3] I used to hate that nickname; after I left middle school, no one called me by that name. A few months ago I met one of my former classmates, Heigoro Iwazaki, after forty years; during those years, we had walked different paths. In middle school, I was relatively short, but Iwzaki-san was tall and, unlike myself, had a lively personality.

[2]Hall of Quiet Thoughts
[3]John Doe

We talked for some time about the old days, about friends and teachers. Suddenly, I was surprised when my old nickname popped out; because I disliked the name so much, I had been trying to forget it. Looking back now, I feel it has an ironic and even nostalgic ring to it. In that name there is the picture of a person who lives in his own world.

When I published my first research, ten years after leaving middle school, I was anxious for the world to verify the validity of my discoveries. If they were correct, I thought, the scientists of the world should confirm them. But that the recognition of my theory should lead to so many problems of various kinds, things that are most destructive to scientific study — that was totally unexpected. It must be admitted that scientific study, in a broader sense, exists for human beings and can always have totally unexpected social consequences. However, if citizens value scientific investigation, I would ask them to leave the scientist in his laboratory, and not to drag him out into the complicated world. This is not only my own opinion; it is probably shared by many scientists, and I am pleading their case.

A long time ago, I ceased to be a no-name *Gonbei;* now no one will leave me alone. I am not unhappy to feel that I am of some worth, but neither can I deny that it is a heavy load for me to bear.

Right now, I am sitting in a chair in my office at the University while I feel the twilight stealing toward the windows, and I am recalling the days when I was an inconspicuous boy. How peaceful it is not to be noticed! For some time, the sound of typing has ceased in the room next door. Those who come to work each day are ready to leave for home. My room has cooled off, the wind must be up. The reflection of the tree branches in the window are restless, and with my mind's eye, I see some boys leaving for home; on my way home from the University, I sometimes see these boys on a windy street corner, or under the gate of a small lonely shrine.

The sun had set; but not for boys at a *kamishibai*.[4]
Far away are the days of childhood.

Idle sorrow, the rueful day has brought back.
The twilight mountains of Kyoto to me.

The Mt. Hiei mirrors mute and departed men.
My thoughts turn to long-parted friends.

I have many memories of the mountains of Kyoto, which I climbed as a boy. Climbing Mt. Yoshida and Mt. Daimonji were really no more than walks. How many different routes I took to climb Mt. Ei: the seven corners of the "White Creek" path which are said to be frequented by bears; the steep inclination of the "Kirara" slope. My heart seemed less troubled when I climbed mountains.

The school had a rabbit hunt from time to time. There were many rabbits on the mountains, Iwakura and Matsugadake: A large net is erected in an appropriate place, while the students form an advancing line and climb the slope, chasing the rabbits out from under the bushes and the long grass. "Hoy! hoy!" the voices echo. A brown rabbit hops like a bouncing ball from under a bush unexpectedly, and the students chase it in the direction of the net. (It should have been a wonderful experience for a middle school student, but after only one such hunt, I came to hate it. That was because dressing the rabbits was such a gory affair.) The upperclassmen grab the rabbits caught in the net, and with machine-like precision, the knees of the animals are broken. The breaking bones make a strangely loud noise in the forest at twilight.

I have a very unpleasant feeling. Looking at the ground, I see small flower petals lying on the grass nearby. Momentarily, I feel relieved; but on looking more closely, I see they are pieces of fur torn from the rabbits. My classmates cheerfully carry the

[4]Old fashioned picture shows, shown at carnivals, etc.

rabbits down; I, alone, am feeling unsatisfied. We are returning to school and it is far from over yet.

On the school grounds, holes are dug to hold fires. The rabbits have been turned into thin red pieces of meat, and some pork has been added to supplement the rabbit meat. Pork and rabbit go into the pot over the red fire, making a thick stew. The sun has finally set, and the students sit around the fire and enjoy the stew. It tastes good to my empty stomach, but I am not happy at heart. The crackling of the fire reminds me of the breaking of the bones. (That is the kind of student I was.)

The inland town of Kyoto is surrounded by mountains on three sides, but learning to swim was mandatory at the school, which owned six boats for this purpose. Students were taught to swim during their first year; lessons were held during the summer at the town of Tsu in Mié Prefecture. This practice seems to have continued long after I left the school. About one hundred students stayed in the main building of the Temple of Kansho-in for three weeks in August. There were big old pine trees standing there, in which the birds made strange sounds all night. It was that kind of temple.

The swimming style was the traditional Japanese way, called *kankai-ryu*. Out motto was, "The sea is like the land." Speed was not important; long-distance swimming was merely a matter of floating in the sea. We were fished out into the boats for thirty minutes at noon and fed hot porridge, but then we went back into the sea. The snack at three o'clock was candy-soup drunk in the sea. It was important to build endurance to stay in the water this long. The standard of swimming at that time was 109 meters the first year, 13.5 kilometers the second year, and 18 kilometers the third year.

A teacher named Tanioka came to supervise the students each year. He was exactly the "daddy" figure that his nickname, "Ototsan" represented. He lived in the administration building inside the school grounds of the First Middle School, and unruly boys sometimes played tricks on his young daughter. For the entire three weeks in Tsu, Tanioka-san stayed on the beach and never entered the water. The students used to say that he practised the true *kankai-ryu*.[5] "If I were in the water, it would be impossible to keep an eye on some one hundred students," seems to have been his reasoning. In any case, it must have taken great patience to spend days on the burning sand watching the boys' heads bobbing on the water. Perhaps it was owing to Tanioka-san that there were no accidents during the several decades that the First Middle School went to the beach in summer.

Sometimes Tanioka-san gave speeches; the one thing he always insisted on was "You are not to go to the Kannon Temple." The students, however, became thirsty at night after a full day spent in the water. After dark, the Kannon Temple in town was lively, and there were ice-shops there, made of walls woven of reeds. To go on "Kannon pilgrimage" meant to steal away to drink the ice-water at these shops. Although the teacher tried to stop us from going, I went with my friends during the first year, and I became sick. That did not stop me, however, from going on that pilgrimage again. Our allowances were kept by the teacher, who supplied one yen per week, out of which we had to pay for supplies used during our study period, when necessary. Drinking ice-water in secret was about the only luxury we could afford.

My eldest brother, Yoshiki, went on those swimming lessons during his middle school years, and so did I. Not only that, but having finished the school in only four years, I went on the trip the next year, and was awarded a certificate as assistant instructor, so that I have numerous memories about the swimming. What I remember most clearly, even now, is the mysterious feeling of swimming far off the beach and suddenly noticing that thick black clouds covered the sky. It was, to be

---

[5]Kankai means "to watch the sea"

sure, a kind of fear, but at the same time it held a bottomless feeling of loneliness.

Let us return to the subject of my studies. I was never really studious; that is, I never rose at four in the morning and studied until breakfast. In fact, I never woke early in the morning, and did not get quickly out of bed when Mother came to wake me. She always used to laugh, saying: "Hidé is *negoi,*" which meant something like "falling into deeper sleep near morning."

My father never forced any of his children to study. He probably wanted each of his children to follow the field of study most suited to each, and he considered that studying only to earn good grades in school was most foolish. During my pre-school years I was afraid of my father, and during my boyhood, I expressed my criticism of him by my silence, but I certainly liked his not goading us to study. In fact, I was never enthusiastic about studying before tests. Furthermore, I do not have a good memory and never did well in "memory courses." Instead, I became more and more interested in mathematics. Umakichi Takenaka, who taught it so well, may have had something to do with that.

I was captivated by the beauty of Euclidean geometry. The clarity, simplicity, and the solid logic of mathematics, especially geometry, attracted me. What delighted me most was that I could solve difficult-seeming problems by my own effort. Geometry taught me the joy of thinking; I became enthusiastic when confronted by problems that would require hours of thinking. I could not hear my mother calling me to dinner. The joy I felt when I finally grasped the key to the solution of a problem gave me something to live for. I solved the problems in the text book far ahead of the class. I bought supplementary books and problem books, and I solved those problems, too. A book called *Understandable Geometry,* by Taketaro Akiyama, was published around that time. It was a most interesting book, and it contained stories about famous Western mathematicians.

I liked algebra, also. In grade-school arithmetic, there is something called stork-and-turtle reckoning.[6] These problems can seem like magic tricks, but they can be solved easily with the use of algebra. All one needs is to write the unknown as x and follow the path of logic. Mathematics suited me, with my simplistic mind that was not satisfied unless everything was completely understood. But looking back, I am glad that I did not become a mathematician. I probably would not have been successful; I will state the reasons later.

Although I was drawn to geometry, I had little interest in physics. The textbooks were simple, and I could understand them easily on first reading. However, what was written there were mere facts, and upon further thought, what lay behind the facts was elusive. This unknown world was distressingly wide, and I could not find a handle to open the door to it. No, I could not figure what or how to think about it. I could not find suitable supplementary books, and the experiments we performed did not satisfy me at all. Perhaps it was natural that I felt no interest.

My worst subjects at school were drafting and physical education. Floor calisthenics were no problem, of course, since I had been active in boating, swimming, and baseball; what bothered me was the horizontal bar. As soon as I entered middle school, I noticed a tall student named Matsuura. Not only was he tall, but he also had the physiognomy of an adult. As I was short and had a childish face, he seemed unapproachable to me. Once Matsuura had mounted himself on the horizontal bar, he did all kinds of tricks. He did the dangerous (so I thought) "wheel" with perfection. I felt dizzy when I saw his tall body rotate around the bar in a beautiful straight line. Naturally, I cowered and hid behind the other students for fear of being called upon by the teacher: "You, Ogawa, it's your turn."

Whenever I had time, I went to "Seishikan" library. I had a

---

[6]These problems concern numbers of storks and turtles. The total number of feet is given, and the number of storks and turtles belonging to the feet are computed without using algebra.

strong desire to read, and it was at the same time a way to retreat into my private world. Often, I read books on European history; on returning home, I would read translations of European novels. I remember being attracted by the prose of an author named Genzaburu Yoshida, whose sentimentalism evoked a curious sympathy in me.

It may have been later that I was moved by Romain Rollands' *Jean Christophe,* and at the same time loved *Sanka-shu* of Seikohoshi. I read the morbid novels of Hakucho Masamune. However, what started me musing was Tolstoi's *On Life.* I read this book during my first years in middle school, but Tolstoi's humanism did not take root in me immediately; even if it had, there would have been no soil to nourish the seed. Still, I cannot deny that it became one of my motives for beginning to think.

The one who first presented me with the question that one must face some time — "What is life?" — was Tolstoi. I can no longer recall what is written in his *On Life,* and I do not even wish to reread it, but I, too, began to think: "What is life?" One of the stages of thought of boyhood is to realize that human beings are plagued by doubts. The next stage is to see one's own doubts clearly; when they are viewed consciously, one realizes that the doubt permeates not only oneself, but all manner of people in the world.

It was not that I did not make friends in school, but when I realized how easy, and how sad, it is for people to hurt one another's feelings, I tried to avoid contact with people. before knowing the loneliness of mankind, I knew my own loneliness. I saw lonely pride in the Hiei Mountain, seen from the Kojin Bridge, on my way to and from school. Perhaps the heart of the beholder is reflected in the mountain. I loved Mt. Hiei for its loneliness. That was the sentimentalism of youth, but curiously, that same sentimentalism remains with me at my fiftieth year.

I had brothers. Because of my oldest sister's marriage, the

number of people in the household decreased, but my brothers and I became more outgoing as we grew older, and we became rather argument-prone. I talked about many things with my brothers. When we disagreed on something, we had arguments; when these could not be resolved, we had fights. My opponent was often my next older brother, Shigeki.

Shigeki knew a lot of things that I did not know. He absorbed new information quickly from friends and teachers. The reason I read Tolstoi's *On Life* was because he tried to confuse me with what he had learned from his friend, who was a Tolstoian. Tolstoi was very influential among the youth of that period; one of my middle school friends later joined a "New Village" organised by Saneatsu Mushakoji. Shigeki's arguments were usually solid, which made me want all the more to disagree with him. I, who was smaller, was always the one who lost the argument and began the fight. This was usually broken up by my oldest brother, Yoshuki, in a rather violent manner. He never tried to talk to us, but walked up to the two of us, held one hand against the backs of each head, and struck them together. That was painful, and I would cry in chagrin. Finally, I would retreat again into myself.

Chapter 6

The Wave and the Wind

Did my temperament finally make me reject the understanding of those around me? My father, Takuji, said that I was the kind of person who always made his own decisions, and he added: ''I cannot even guess what Hideki is thinking about.'' My brothers were probably more open than I was. Tamaki, who was my junior, had a straightforward personality and was the most trusted in the family. I was the middle one of the five brothers; I was the valley in which no light could enter; and what cold wind blew, or what rivers ran in that valley, even my father could not guess. The saying: ''No one knows the child better than his father,'' my oldest brother, Yoshiki, once said, ''does not apply to Hideki and our father.'' Perhaps that was so.

My brothers and I had the habit of reading in bed. When I was tired of reading, my thoughts sometimes ran to the future, or sometimes to the past when I was a very small boy. At those times, a particular scene was apt to recur, bringing back the feeling of exaltation that I got when the gas lamps were first installed on Kawaramachi Street.

That happened before I entered elementary school. Was it autumn? The wind was blowing through the street and a fine rain soaked my collar. Mt. Ei was already greyish in color, and night seemed to be descending its slope. I was standing before a shop, and I cannot remember if Mother was with me or not. People hurried in front of me on their way home. In the bustle of the evening, my ears caught the voices of children and I turned toward them to see a tall man walking toward me, surrounded by a group of ten or so jubilant children. He was carrying a small ladder and a long pole; from one shoulder hung a can.

This group stopped about a dozen feet away, where the gas street lamp stood, erected only a few months before. "Can I do it?" a girl's voice asked. After that, the boys tried to take the pole from the man's hand. He calmed the children and then extended the pole toward the lamp. A blue flame leapt up, and the fine rain made a glowing sphere for a meter or so around the lamp. The falling rain glittered in the light, and a cheer arose from the children. Then the little group, with the man in the center moved toward the next lamp. The lighting of the lamps, one by one, in the twilight town was a mystical ceremony in the eyes of the child. How far was this man going to go? He was a magician; and it seemed to me that an unknown world was being revealed that existed beyond the row of lamps. That memory is now far away, but the seed of pessimism, meanwhile, was sown within me.

I avoided my father; I fought with my brothers. I had few intimate friends, but it was not true that I was close to no one. I was the favorite of both my father's mother and my mother's parents, who lived with us; I was always close to my mother; my youngest brother, Masuki, was also very dear to me. I remember with fondness the time when, of all the children, only Masu-chan remained at home, after Tamaki began to attend elementary school.

Masu-chan eagerly awaits my return from school when I open the front door and announce: "I'm back." Masu-chan runs in delight to open the *shoji* screen; meanwhile, I go to the kitchen entrance. Masu-chan, seeing that I am not at the front door, goes into the kitchen, but I am already again at the front door. I used to play these affectionate tricks on my youngest brother; I loved him so much. However, this youngest brother left this world before me, as a casualty of the last war.

I did a fair amount of service for my sisters, also. When my older sister Kayoko, who married and lived in Tokyo, had her first child, my second sister, Taéko, still living with us at that time, went to Tokyo to be helpful around the house. It must have been during a school break, and I was to accompany her, not having been to Tokyo since my infancy there. We took two rickshaws through the quiet town of Kyoto, toward the railway station, my heart full of joy. Arriving in Tokyo, we went to my sister's house in Aoyama, inside the city. My younger sister was busy helping, but I had nothing to do. I looked through my brother-in-law's library, took walks around the neighborhood, and relaxed. The days passed quickly, and soon it was time to go back to Kyoto. In due course, my second sister also got married. While she was living at home with us, her school friends sometimes came to visit, and cheered up the house. When only my brothers were left, the house suddenly seemed barren.

My father was a man of unpredictable moods. Sometimes he came home from the University after we had all finished dinner. Then he would enjoy his own dinner with a little saké, while we all sat around him. He thoroughly enjoyed his food, and in this good humor he would tell interesting stories.

My father used to have many visitors, and he was a man who told everyone exactly what he thought. It was always clear weather when his audience listened quietly, but when the listeners said things not to his liking, the weather quickly deteriorated. "What the devil do you mean by that?" Father's loud voice could be heard in our rooms and made me nervous, even though I was used to it.

Was it because of my father's stormy temperament that mine took the opposite direction? Or did I consciously react against the Confucian thought that had surrounded me since infancy? For some time, I had been seeking something that was to be found in the writings of Lao-tzu and Chuang-tzu. What I had learned until then in my house were the Confucian classics in the main stream of Chinese philosophy. My brothers probably did not feel too much opposed to this, but to me, Confucianism seemed an unnatural philosophy. It was something imposed upon me before I had any ability to criticize, and this very fact made me suspicious of it.

"One's body is received from one's parents. To inflict hurt upon it is ..." The very style seems to impose the thought upon us. No deep philosophy appeared to be behind it. I began to look for something less dogmatic. I read the *Doctrine of the Mean* in my father's study; but this was somewhat *too* philosophical. I wondered why my father had not required me to study that. And then I found, first Lao-tzu, and afterwards, Chuang-tzu.

My boyhood pessimism was deepened by my reading of these books. They did not merely support my desire to disagree with what I had been taught, but contained something genuinely appealing to me. More than ever, I withdrew into myself, although there was no definite occurrence, no one event that particularly depressed me. I did not fall in love; in those days, relationships between the sons of respectable families and girls were taboo.

After I began to delve into physics, I still felt a desperate pessimism when my work was not going well. I have learned since then that a number of European physicists committed suicide, and I feel that I can understand that. However, I myself have never thought of committing suicide. Inside me there is a sense of responsibility to man, to society, to the family that is a building block of society, to my friends, and to young researchers. This seems to exist independently of my abhorrence for the emptiness of man and for the contradictions within society. It is not a matter of "give and take," but rather a sense of duty, to give without taking. An unrewarded good deed may be related to the *mu* (nothing) of Lao-tzu and Chuang-tzu.

Faith in science provided me with some relief from my depression, but at the same time, it revealed new evidence to support pessimism in the scientific view of nature. What always supported me under this pyschological stress was the possibility of continuing my creative activities. Without that, I would be holding a hand of cards without any trumps; that is why I cling on so to theoretical physics. I am, perhaps, unconsciously seeking a harmony and simplicity beyond human problems and contradictions. To me, when I was attracted to Lao-Chuang, even the liberalism of Sotosaburo Mori, whom I admired, seemed more Taoist than European.

Before I became so taken up with Lao-Chuang, I belonged to a small literary group, connected with the school magazine called *Konoé*. About two years before I entered the First Middle School, the magazine had been started by a group which included my brother Shigeki and Takeo Kuwabara. Its name was that of the street that ran in front of the school. Some tens of members handed in manuscripts: stories, essays, themes. There were criticisms of the teachers and proposals for the school. The table of contents usually listed twenty or thirty names, although they were all pen names. Students seemed to write whatever they felt, with the recklessness of youth. That kind of freedom was possible in the First Middle School.

The magazine was started by my brother's class, two years ahead of mine, but it was inherited by my class. The one in-between did not have the temperament to join *Konoé,* so skipping a year, our class was drafted. I do not know how many issues eventually appeared, and I understand that when the magazine disbanded, each of the main contributors kept one old volume; it would be interesting to bring them together.

The cover was drawn by an artistic member of the group, and there was a table of contents with illustrations. The manuscripts were handwritten by the authors, and were bound in the Japanese style, using Japanese paper. Most of it was written with the brush, but some parts were in pen or pencil; the first page of each contribution was usually illustrated by the author. The magazine was circulated among the members, who wrote criticisms in the margins, some of them most interesting.

I wrote a children's story about encouragement by friends. I do not remember very well what it was about, and I wonder how I would feel if I looked at it now. But regardless of its contents, it is noteworthy for me to know that there was a time when I could write a children's tale. Indeed, I have always believed that literary beauty is not so far removed from the "beauty" that is revealed to us by theoretical physics. I even retain a desire now to write a children's tale if I can find the time.

Kontaro Kawasaki, one of my close friends, was especially involved in editing the *Konoé*. The editors, having collected the manuscripts, held an editorial meeting. Probably, it consisted only of an exchange of boyish, sometimes selfish, opinions. "Let's put this one in front," or "This one is not really very good." The order was decided; the table of contents was constructed. They must have felt like real editors then; I myself had almost nothing to do with the editing.

The philosophies of Lao-tzu and Chuang-tzu were naturalistic and deterministic, and they had a kind of solidity which must have been one of the aspects of it that appealed to me. I could never be satisfied with half-way thinking, even from the time when I was quite small. As I said before, every day at the elementary school there was a morning ceremony, at which the principal, Tatebé-san, made a short speech. I have forgotten most of the speeches, but curiously I remember one of them very clearly. One morning he spoke of "solidity": many animals cross a river, all of them swimming except the elephant, who walks across on the bottom. That was "solidity". The word left a lasting impression on one elementary school student, who wondered childishly what would happen if the river were too deep for even an elephant to ford.

In middle school, I enjoyed the paradoxes of Lao-tzu and Chuang-tzu, but I also could not deny that there had to be something more to existence. The blood of youth was beginning to flow inside me. I believe it was in the fourth year of middle school that I learned the fundamentals of the theory of evolution. It was taught by the head teacher, Ushinosuke Takeda, whose nickname was, "Ushiyan."[1] He was an elderly man and very erudite, but also a good lecturer. First, he introduced the theory of Lamarck: when an organism uses a particular part of its body continuously, that part becomes highly developed and that is how organisms evolve. That seemed very logical to me, but Takeda-san said that theory was wrong. Characteristics that an organism acquires after its birth are not passed down the generations, so they are of no use to evolution.

Next came the explanation of Darwin's theory, based on competition to survive among the same species of organisms. Those who win, breed more successfully. It is through the survival of the fittest through natural selection, that organisms evolve. I could not understand this clearly; it bothered me when I went home, and I walked about the garden, thinking about what I had been told.

By this time, we were no longer living in the house on Kawaramachi. Viscount Toyōka had sold both his own house and our rented house and had retired to Nishikamo; we never heard from him after that. I believe our house was bought by the industrialist, Gondo Yamaguchi. Because the house that Toyōka-san had lived in was larger, we rented that one and moved into it. Soon after that, Kawaramachi Street was widened, and a tram began to run on it. At the same time, the land along the street was divided up into small lots and some stores were built. Although the world seemed very quiet around

[1]Mr Cow

the tenth year of Taisho, it was really changing rapidly.

Part of our backyard was a bamboo forest. Opposite the kitchen were chicken coops made of bamboo. My mother had begun to keep some Leghorn hens to provide eggs for the children. In the relatively wide central area, we played catch and shot put. In the corner, near the chicken coops, was a horizontal bar, but we did not use it very much.

Having just heard about evolution, I did not even think about catch or shot put that day. Instead, I walked back and forth in the garden, thinking about the mystery of evolution. In order for natural selection to occur, there must be a difference between the fit and the unfit from the very moment of their births. If one took into account the differences that might occur after birth, that would be no different than Lamarck's theory. Then where did the inborn differences come from? Takéda-san's explanations on this point were not very clear. No matter how it had been explained, there was no way that I could understand the theory of evolution with the meagre knowledge of a middle school student.

Thinking about it much later, I realize that at that time I had a firm belief in the necessity and inevitability of each event in nature. That was probably why I found it easy to accept determinism as a logical system of thought. But a phenomenon called evolution, that could be seen as purposeful, happens to an *entire species* of organisms. I must have felt, in some childish way, that some problem was there that could not be resolved by my own simple thinking.

At that time, I was not aware that there had been a revolution in physics at the beginning of the twentieth century. I had not even fully realized that what I believed to be the only logical system of thought was in fact what scientists, until the end of the nineteenth century, believed to be absolutely correct. Of course, I had no notion that new theories like the quantum theory and relativity had arisen at the beginning of the twentieth century and were shaking the foundations of classical physics. However, the fact that I had trouble understanding Darwin's theory of evolution seems to have had a deep effect upon my psychological growth. My unconscious, at that time, began to respond actively in a direction different from before.

My boyhood sentimentality seems to have been changing into the romanticism of youth. Thinking back, I can see that there was a reason, after all, for me to be attracted to Chuang-tzu, who seems to be intoxicated with his own magnificent imagination, rather than Lao-tzu, who represents the wisdom of those who know life's emptiness. The blood of youth began to flow in my body, and I was eager to leave the First Middle School and to enter the Third High School next door. It was about this time, close to the end of the second semester of my fourth year in middle school, that Albert Einstein visited Japan.

Writing a children's tale in the school magazine *Konoé,* attracted by the beauty of Euclidean geometry, troubled by the theory of evolution, approaching life's meaning through the writings of Lao-tzu and Chuang-tzu — the erratic course of my boyhood sentiment now draws a smile to my lips. I was like a ship whose radar is seeking landmarks in the dark. My boy's radar finally succeeded in finding something, but it was far far away, so far that I could not tell what meaning it held for me. Only after it had drifted out of range again, did I realize its significance.

At the end of summer, the eleventh year of Taisho (1922), the newspaper announced: "Dr. Einstein visits Japan." I was in the fourth year of middle school. Among the Japanese, Einstein's theory had been discussed, without being understood, although Jun Ishiwara had promoted this theory from its earliest days. As soon as the forthcoming visit was noticed, Einstein's name flooded the newspaper and magazines. His trip had been arranged by the publisher of the magazine *Kaizo-sha* which disappeared some years ago from the publishing scene, but at that time it shared half of intellectual Japan (along with

Chuōkōron-sha), and must have been prosperous.

Dr. Einstein arrived in Kobé on November 11, 1922, aboard the ship *Kitano Maru*; he had been awarded the Nobel Prize half a month earlier. The great figures of the Japanese scientific world, including Hantaro Nagaoka, Aoya Kuwaki, and Jun Ishiwara, welcomed him at Kobé Bay. Arriving in the afternoon, Einstein went directly to a hotel in Kyoto, where he spent the night. A few days earlier, the magazine *Kaizo* had released its December edition, a huge "Einstein issue," with Einstein's picture in front and with fifteen articles on Einstein and his work by the scientists mentioned, as well as Torahiko Terada, Kinnosuke Ogura, and others. Among them, I cannot fail to mention, was Kijuro Tamaki, a professor at Kyoto University whose help I received from the time I chose to do theoretical physics. At the time of Einstein's visit, however, I did not know Professor Tamaki's name.

After one night in Kyoto, Einstein went directly to Tokyo, accompanied by the scientists who had met him on his arrival in Kobé. From then on, pictures and stories about him turned up almost daily in the newspapers. His first lecture, at Keiō University, was said to have lasted five hours (including an intermission). Although Einstein had specified an audience not larger than one thousand, the crowd was twice as large, although there were only a few Japanese scientists whose specialty was physics.

At the time of Einstein's visit to Japan, my interest in physics was not great; instead, I was involved with mathematics. It was unusual for the work of a physicist to be known outside of a specialized circle, but books on philosophy began to multiply; the Iwanami Press philosophy series already contained many volumes. Ikutaro Nishida's *Study of the Good* had been inspiring young people for years, and it was probably at about this time that Dr. H. Tanabé's *Recent Natural Sciences* was issued as part of the philosophy series. In this book, "quantum theory" occurred several times. I did not understand its meaning at all, but I felt a mystical attraction toward the words, and I began to admire Max Planck, the originator of the quantum theory.

Physicists generally had an existence entirely separate from the world of journalism, but Jun Ishiwara was an exception. He wrote frequently in magazines and newspapers about the new directions in theoretical physics, and also wrote a book called *The Principle of Relativity*. I had heard the name of Albert Einstein. Unconsciously, I might have been moving slowly in the direction of theoretical physics, for I became involved in physics experiments in my fourth year at middle school. The experiments were done by teams of two students; my partner was Shinichiro Kudo, and we did an experiment to measure humidity.

When ether is suddenly expanded, it lowers its temperature by evaporation; beads of water condense on the surface of its metal container. By comparing the temperature of the container with the temperature of the room, the humidity of the air can be calculated. I succeeded in this experiment and was quite happy about it. Kudo said suddenly: "You will become like Einstein." At that moment, I had no idea what he was talking about, as I had no thought of becoming a physicist. However, after the experiment was over, I became very happy for some reason unknown to me. I was in the kind of confused state mentioned by Chuang-tzu.

Dr. Einstein was a great figure, who was very remote from me. Kudo's words did not seem to apply, and yet his words appear to have made an invisible crack in the icefloe that blocked my ship. The French poet, Proudhon, had written: "A vase was struck lightly with a fan, which did not leave a visible mark. The crack grew with time, and one day the vase broke by itself." As I wrote the words just above, this passage entered my head.

Einstein went from Tokyo to Sendai, then back to Tokyo, and returned to Kyoto in December. He lectured there before a full house, although Kyoto was a place where people seldom gathered in groups. The lecture was on that difficult subject called the Principle of Relativity. Perhaps even the people of Kyoto, who usually showed no quick affection, were attracted by Einstein's personality; or it may be their peculiar quality of seizing upon new things that brought them out this time. No, it cannot be those things. The Theory of Relativity and its originator had been a topic of conversation in all the civilized countries of the world, and neither Japan nor Kyoto could be exceptions.

However, these are things I discovered and thought about later. At the time, I did not even comprehend the words of my friend Kudo. When Einstein lectured in Kyoto, I did not attend, and I did not even know when or where the lecture would take place. Akira Kobori, who became my classmate in high school, and later became a mathematician, heard the lecture. Why was I so thoughtless? In a few words, not only was I uninterested in what was happening outside my own little world, I did not even know who I was, and what changes were occurring within myself.

Einstein soon afterwards left Japan and returned to Europe. Some sixteen years later, when I was thirty-two, I left for a foreign country for the first time in my life, and was in Europe, having been invited to a conference in Brussels to speak about the meson theory. Relativity was no longer at the center of theoretical physics, and elementary particle physics was beginning to take the stage. Einstein, himself, had left Europe. But fate was perverse. World War II was imminent and I decided to leave Berlin, where I was visiting, just before the outbreak of hostilities. All international meetings were postponed indefinitely. I boarded the *Yasukumi Maru*, which was returning to Japan by way of America. I stopped at New York and went to Princeton to visit Dr. Einstein, who already had silver hair.

After the war, I visited Einstein many times at Princeton, and my admiration for him increased with the years. During recent years, I have been studying how to apply the great concepts of Einstein, in the General Theory of Relativity, to the world of subnuclear particles. I would like to see him once again; but now it is impossible.

Chapter 7

Episode

If you had just walked across a meadow and then were told that it contained a pitfall, a deep one with no way out, you would shudder. Moreover, that pitfall was covered with grass and would be unnoticeable even if you walked right up to it! My rather ordinary life seems to have contained such a danger, but I learned about its existence some thirty years later. Thus, it had little impact upon me, and still seems not quite believable. My mother told the story to an acquaintance, so I cannot simply say it is untrue. In a way, this episode may even add some depth to my life's story.

At the time, I knew nothing about it. I walked upon the grassy meadow and felt the heat of the sun reflected from the grass, and did not know that there was a pitfall there. That is why I cannot continue to write about this incident with "I," and will adopt the third person (unusual as that may be in an autobiography). It will also help me to be objective.

The time was long before Einstein's visit to Japan. At Takuji Ogawa's house, his father-in-law Komakitsu had passed away; the eldest daughter, Kayoko, had left; the second daughter, Takéo, had also married and moved to Tokyo, and only the boys were left at home. The eldest of the boys, Yoshiki, then at the Third High School, was planning to attend Tokyo University. Then another child would leave the house and it would become still more lonesome. After the second son, Shigeki, had left, what were the plans for Hideki, Tamaki, and Masuki?

Takuji considered the distinctive features and strong points of his children. His friends often told him: "You have good children. They are all exceptional." When he heard these judgements, he would imagine a time when all were grown up and each had his own family, and a smile broke out. But it was difficult to see a child through his career, and he was not without anxieties.

Takuji and his wife had decided to raise all of their children to become scholars. Did they all have the potential, though? Even if they did, that would require a lot of money. The income of a professor in Japan has not been very extravagant in any period of history. It was not easy for a professor to arrange marriages for two daughters and put five sons through the University. Takuji had decided to lead his children into lives of scholarship, and he was proud of that. But then, he began to wonder whether there were other respectable professions, and he suddenly decided to re-evaluate his children.

This train of thought appeared in Takuji Ogawa as suddenly as the wind, and it left a long trail behind. He had given his all to scholarship and knew the happiness and sadness of a scholar's life. It was impossible for him even to imagine that there could be days in his life when he would not function as a scholar. But that is not all of life; man has many ways to live. Perhaps it was insolent of him to plan for his children only from his own point of view. It might be better to have one of his sons in a different world, maybe more natural. If so, which of his children should it be?

His third son, Hideki, came into his mind. He was certainly different from the other children! He had childish features, and sensitive and delicate feelings. What was behind the exterior? Was his a soul that was superior to the others, or ...? Takuji shook his head. That child was suppressing something inside, but when that something surfaced, he appeared to be too independent. He had the most difficult character of the five boys, and sometimes that worried him. Takuji, in fact, had said: "I can never tell what Hideki is thinking about!" Perhaps he was mistaken to think that Hideki should also be a scholar. If it was a mistake, he must think of a different life for him.

It was afternoon, the children were all at school; Takuji was sitting at his desk with a book before him. His wife entered quietly, and said: "Don't you have to leave now?"

He answered: "Yes, it is about time."

Takuji looked at his wife, and then turned away to face the garden. It was the time of year when the leaves were deepening their green. The shrubbery smelt of new leaves, and the yellow roses were brilliant.

"How is Hideki these days?"

Hearing no answer, he turned again to his wife and noticed that her eyes were staring in wonderment. Perhaps she was surprised at the strange question; she answered, "Nothing really different."

Takuji at this moment realized that this was going to be a rather serious matter. His wife seemed to sense that something was up; her eyes betrayed her anxiety. Since he had begun, however, Takuji continued: "Is he planning to go to the University?"

Takuji saw his wife turn pale as she heard his words. "What do you mean? I don't ..." and as she stammered, he finished for her " ... understand." He stood up, entered the adjoining room, and began to remove his kimono. If they kept on in this manner, he was afraid the conversation might turn grim. His wife helped him with his shirt, handed him his tie. Takuji did not go on with the conversation until he was fully dressed. He had not made up his mind; rather, he was thinking aloud and wanted to try out his thoughts on his wife. Though she had asked what he meant, she knew perfectly well. That was why she changed color when he had spoken. A problem that is important to a child is always important to his mother. Besides, Hideki had a strong resemblance to his mother in his silence and his quiet nature. They were alike, too, in that their quietness concealed an inner strength.

Having finished dressing, Takuji returned to his study to check the contents of his briefcase, while he wondered: "Has my question already shocked her?"

"Are you leaving?" his wife's manner was serious. "What about Hideki?"

Takuji turned and smiled a smile of tenderness for his wife, who had shared for so many years his joys and sorrows. "We will talk about it tonight."

She nodded.

"Think about it until then."

"Yes."

She accompanied her husband to the front gate, and as she handed the briefcase to her husband, after he had finished putting on his shoes, she said: "I think Hideki should go to University, also." No answer. "What made you think of such a thing only for him?" No answer. "There are children who are not conspicuous. Not only those children who stand out and seem talented become men of great achievement. Those who are inconspicuous as children often ..."

Takuji heard his wife's serious words, and her voice reminded him of her birth and upbringing. Furthermore, he felt her pride and strength as a mother. As her mouth closed after each telling phrase, it prepared for the next group of proud words: " ... and we should do the same for all our children. We cannot be unfair to one of them."

"All right, I understand your view. Let's talk about it tonight," and he left.

And he thought: what his wife had said was true, and fine in its way, but his own thoughts were not wrong, either. It was the duty of parents to let children find the ways best suited to themselves. The Ogawa family had unconsciously directed the children to become scholars. Without doubting at all, they had thought only of raising them to be scholars, and the children, too, had probably accepted this idea long ago. But now the children, one by one, were moving from childhood to youth, and they were becoming more aware of their own individualities. Was it right that they should all be directed along a single path? Do parents have that right?

For some time, Takuji had been considering sending his third son, Hideki, to a technical college. Hideki was not as outstanding a student as his older brothers, and for that reason Takuji thought it might be better to send him on a different path. That was not unfair, for even if his five sons were to take five different paths, it would still be fair, provided each path was appropriate. Indeed, it would be unfair to force each one to walk the same path, whether he liked it or not.

On that day, Takuji observed, he spent more time than usual thinking about the children. Indeed, with every passing day, he spent more time thinking about the children. He could see them approaching their crossroads, about to enter crucial stages in their lives. If he simply let them alone, however highly he regarded their judgements and their intelligence, he would be failing in his duty as a parent.

The Ogawas did not have the chance to discuss Hideki's future that night. They felt instinctively that it was not a thing to discuss at a time especially set apart for that purpose. It was possible that it might create an unusual bitterness between them if each had not thought the problem through to its limits. The difference would be all the more difficult to resolve because each of them had only the best of intentions.

One day, leaving his office at twilight, noting the old-fashioned color of the red bricks, Takuji walked through the poplar trees and came to the edge of the campus. Someone called after him: "Ogawa-san!" Takuji turned and saw the principal of the First Middle School, Sotosaburo Mori.

"Are you on your way home? Busy as always?" They began to walk together, and Takuji said: "My children are always. giving you trouble."

"No! no!" denied Mori. "They are all good children."

Takuji examined Mori's face; his voice was clear and straight-forward. Something flashed inside Takuji: It would be a good idea to talk to this man.

They walked some steps in silence, and then Takuji spoke while he looked at the houses before him. "Do you know my son, Hideki, well?"

"Yes, very well."

A brightly colored train slid slowly past their eyes. The white walls reflected the evening sun and were bright. The trees on campus were quiet and their leaves glittered. Students moved past them; some, who had listened to Takuji's lectures tipped their caps in respect.

"I am a little concerned as to which direction to send him."

"What do you mean by 'direction'?"

"Whether I should send him to University after high school or ... (silence) ... or perhaps to a technical school."

Mori did not answer immediately, but looked up at the sky. A line of clouds colored by the twilight appeared like a brush stroke upon the pale blue sky.

"Ogawa-san," said Mori, finally. "I do not understand why you should say that. A boy with Hideki's potential is very rare."

Takuji started to speak, but Mori interrupted, "Wait. If you think I am merely flattering you, let me adopt the boy. I have taught mathematics to him, and his brain works in leaps. His conceptions are sharp and unusual. I do not know about the other subjects, except what I have seen on the grade reports, but as far as mathematics is concerned, perhaps you will not like the word, but there is something of the genius in him. I will stand by that. In the future, he has a high potential but I cannot believe that you do not already know this."

Takuji looked up at the clouds that now burned like a flame, flowing with beauty, and he was saying in his heart, "I knew it."

Chapter 8

Youth

If one considers its earliest foundation in Osaka (after which it changed its name many times), the Third High School is just as old as the First Middle School. The school moved to Kyoto in the eighteenth year of Meiji, and it took the name Third High School in the twenty-seventh year of Meiji (1894). The high school was just to the north of the First Middle School, and long before that school, it had the motto, "Freedom." So both literally and psychologically, to enter this school was merely a matter of moving next door. The entrance examination was not a problem; I had no worry about it as I did very well in the mathematics questions.

When I entered the Third High, in April of the twelfth year of Taisho, I was sixteen years of age. The principal of the school was Sotosaburu Mori, who had been the principal of the First Middle School. That came about in the following way: The year before, when I was in the fourth year of the middle school, there had been a students' strike at the Third High. A new principal, Mr. Kaneko, had fired the old teachers, which was considered an inhumane act, even though Kaneko-san might have had reason for doing so. Some wondered if he was not out to change the "Freedom" motto for something else. The students drew together under the slogan, "for our teachers," with the alumni backing them. The community also was sympathetic to the students and the teachers who had been discharged.

Shigeki, my older brother, had just entered the high school. He decided to stay in a dormitory with the upperclassmen, and so he did not come home that night. My parents were worried, and my father went to the Third High late at night. For some reason I went with him, and we were standing in front of the gate. However, the gate was not opened, and my father had a discussion through the gate with a representative of the students. Under the pale light of the lamp, my father's face was hard. I do not remember what words were exchanged, and perhaps I was not even listening. Apparently I had not thought about the meaning of the strike, but had merely accompanied my father. How impenetrable I was!

My father and I were not able to see my brother that night, and the strike soon ended in unconditional victory for the students; no one was punished. During that summer, Kaneko-san was transferred, as the students had demanded and afterwards Mori-san came from the First Middle School to be principal. In the spring of next year, the storm had passed; I entered the school and enrolled in the *Ko* program of natural sciences.

In the high schools of that era, the natural science curriculum was divided into *Ko* and *Otsu* programs. In *Ko*, the first foreign language was English and the second was German. One of its subjects was "dynamics," and there were no biology laboratories. In that way, it connected with the physical science departments at the universities, especially mathematics, physics, chemistry, and also with engineering. *Otsu* had German as the main foreign language and had biology laboratories replacing "dynamics." Thus, it was more related to biology, medicine, and the agricultural sciences. When I chose the *Ko* of the natural sciences, it meant that I had decided to specialize in one of the sciences other than biology.

Still, youth came before study. While listening to Mori-san's short introductory speech, I was happy. The pessimism of my middle school years was, it seemed, in hiding. After the ceremony, I passed through a hallway whose walls were pasted with posters proclaiming "Baseball Club," "Track Club," etc. Under each poster stood several upperclassmen who were trying to recruit new club members. Afraid of being dragooned into a sports club, I walked quickly through the hall. I had no desire at all to join anything, but unexpectedly my name was called. I was startled, and when I turned and recognized the face, I was puzzled. It was my eldest brother's friend, Katsuho Yoshie, who sometimes came to our house. "You are Yoshiki's brother, aren't you? Why don't you join the Judo Club?"

Now, *judo* was not my favorite sport! Although I had been good at *sumo* wrestling in elementary school, I had been poor in *judo*, which we practised in middle school. I was easily pinned, and being small, I could not get up, once I was down, and had no choice but to surrender. During my stay at middle school, I gave up *judo* and took up *kendo* fencing, with wooden swords, as I was required to take one or the other. So I especially did not want to join the Judo Club. I replied quickly that I had beriberi, and the matter was not pursued. (It was not a complete lie, as I did have a slight case and was taking vitamins as a cure.) I quickly escaped from that hallway.

Although I joined none of the sports clubs, the atmosphere did not allow for studying at the beginning of the school year. For a little while, everyone was unmindful of the classes.

> The hills burn in crimson blossom,
> The green smell of the color of the beach,
> Singing to the flowers of the Emperor's city,
> The moon over Mt. Yoshida ...

We were busy learning the numerous school songs, and each one made our hearts swell with pride. We seemed to be in the very heat of youth, as we sang loudly before Mt. Yoshida.

Did my temperament change then, during my years spent at
the Third High? Its environment was even freer and more
cheerful than that of The First Middle School. For a while, I
also seemed to move in the whirlwind of youth, participating in
the common temperament of high school students of that time.
No, I probably still appeared somewhat different from the other
students. When I entered the University three years later, I
wondered: What did I do in the Third High? And I felt a twinge
of regret at having wasted three years. On the one hand, I
regretted that for three years I had not studied seriously. On the
other hand, I felt that I had not enjoyed my youth as fully as the
others had. Those two feelings were contradictory, but this
realization came later on.

Before I could become used to the new life there, the Anniversary Festival of the school, which fell on May 1, was getting
close, and the whole school became involved in the preparations. For a long time, the city of Kyoto had been very kind to
students, and the Third High School students were especially
favored. I found out later that there was an expression used at
that time, whose meaning was, "Pay when you are important."
For example, a group of students might owe money to a
restaurant, and the owner would say: "Pay it when you become
rich and famous." The people of the city, in general, had
inordinate respect for the students and their intelligence.
Perhaps the Kyoto tradition of placing high value on man's
knowledge turned into simple affection for the students of
the Third High. At the Anniversary Festival, the students
presented their youth to the people of Kyoto, who eagerly
awaited the day. Gaily dressed young women (in Third High
slang, "mechen," i.e., "mädchen") would gather in colorful
groups. To entertain the ladies and gentlemen, each class set up
a bedraggled food stand in a tent, serving coffee, soft drinks,
and some food. The profits were used to pay for the costume
parade and other entertainments. Tickets were the currency
used at the food stands, and each student was responsible for
selling a certain number of tickets in advance.

One sunny afternoon at the end of April, I went with several other students to the main street, all of us determined to sell our tickets. Since they were only festival tickets, it was not as serious a matter as the part time work of today's students. However, passers-by did not willingly buy the tickets. I followed my companions, not having the courage to approach anyone, whom I did not know. I simply did not have the brashness required to sell tickets and then boast of my skill as a salesman. I only walked along, not really expecting anything to happen. I was not a fop, nor a semi-delinquent student, and while I enjoyed being a Third High student, I could not bring myself to accept all of youth's joys. I did not reject my friends' actions, but I was never a leader in them. As for the tickets, I thought that if I asked my mother, she would take care of them. I wanted to go home, and not walk around the town selling tickets.

We walked past the Heian Shrine to the park in Okazaki. There was a baseball game, and some people were watching. My friends walked around the baseball field, calling to the spectators, but when not even a single ticket was sold, they began to feel frustrated. It was then that a young woman approached me, probably having heard the calls of my friends. Looking at the pieces of paper in my hands, she asked: "What are those?" I suddenly brightened: "Tickets for the Anniversary Festival."

"I'll buy some." She took out her purse, while I handed her several tickets with my friends staring in disbelief, their expressions indicating surprise that I was actually selling tickets! But I could not be proud of myself even at times like these. I felt guilty towards my friends as well as towards the person who had bought the tickets, and I thought of the words, "unearned income." Because I felt that way, I never again went out to sell tickets.

The preparations for the festival were proceeding smoothly. Some of the dormitories decorated every room, the theme usually being social satire. On the first of May, during the first festival that I experienced as a student, the school was over-flowing with visitors, and the dormitories were especially crowded and hard to enter. Having finally made one's way into the dormitory, one heard laughing voices, surprised voices, and critical voices, all at the same time. Some rooms were totally darkened, and people waited there tensely; suddenly, an eerie skeleton appeared, and some young women began to scream.

Outside the dormitory there were a number of food stands, with student-attendants busily serving. Meanwhile, costume parades took place on the school grounds. There were some costumes with satiric intention, but many were quite childish. Some, like "The Ogre-hunting on Mt. Oe," seemed almost like something from an elementary school play. An ogre in a large mask walked around the grounds with maidens that he had captured.

Our class chose to do a play in the old storytelling tradition (*kodan*). Some of us were swordsmen; one was the third *shogun,* Iémitsu. Several female attendants were needed, and for this purpose female impersonators were chosen, who put on wigs, whitened their faces, and walked modestly around the grounds, trailing their long kimono behind them. The costumes were all rented from a costume shop. My friend since middle school, Kontaro Kawasaki, was one of those in the parade as a female attendant. I was asked to do the same, but I refused, saying that it was out of the question. It is interesting, looking back, to consider why I thought it was out of the question. Not only was I innocent and unworldly, but also I could not fit easily into this atmosphere because of my family background. Instead, I took the unimportant role of carrying the flag.

The day and its festivities passed quickly, and the guests gradually dispersed. Mt. Yoshida fell into twilight:

> Leaves of cherry trees blow in the winds.
> The clear twilight of the May sky,

Over the thirty-six peaks of the mountain range ...

The students sang the song of the festival, and danced in the fields. It was an explosion of youth, an exhilarating leap of young burning life. I, too, danced from one end of the field to the other that day with my classmates. That day I learned the taste of beer.

Finally the festival was over, but we still paid no heed to studying. There was to be the athletic competition with the First High School (in Tokyo) during the summer vacation. The cheering squad was already organized, its head being Katsuho Yoshié, the one who had tried to persuade me to join the Judo Club. In my own class, a friend was very enthusiastic, and so I was also made a member of the cheering squad. The organizers of the cheering squad waited at the school gate to get students leaving the school to cheer the baseball team. Each day about ten students got together, striking a huge drum and waving red flags to encourage the team members. Nearby, some non-student spectators would be standing around.

The First High-Third High athletic competitions caused an intense excitement that is difficult to imagine now. Youth seemed to have crystallized around it. Drums were borrowed from neighborhood shrines. Our class, to find a suitable drum, went finally to Sakomoto, on the other side of Mt. Hiei. About fifteen classmates took turns carrying it, and when I got home in the evening, my shoulders were aching.

Baseball practice, and the cheering that went with it became more extended every day. The practice finally ended when the early summer sun was gone and one could no longer see the baseball. At about this time, the Third High cheering squad stopped using the red flag, after warning by the police. (The number three was drawn in white on the red flag.)

The most important First High-Third High competition event was baseball. There were also track, tennis, boating, and various other sports, but the school that lost in baseball was required to travel next year to the other school for competitions; thus, baseball was most important. The year I entered the Third High was one of the years we went to Tokyo. As the summer approached, there began to appear posters with propaganda. They were not only for the competition with the First High; as though they were flashes of youth and joy, they carried various messages. But as the expedition to Tokyo neared, the number of posters began to increase. "Destroy the Eastern (i.e., Tokyo) barbarians," or else. "Drunk on the flowers of the East Mountain in spring, and lost in the scent of Kiyo Valley in autumn."

We went to Tokyo at the end of August, occupying several cars of the night train. How many hundreds were we? The importance of this event was measured by the fact that so many athletes and cheering squad members went. There were several hundreds of small and large flags and tens of large drums, and other luggage that flowed out into the passageways. Some students were sitting on mats on the floor. A song would start from one corner and immediately fill the whole car, and then the next car would begin singing. Drums beat the rhythms. In the midnight train that glowed darkly red with the lowered lamps, the voices of the youths filled, spilled, and seemed to echo out of the open windows to the far meadows.

It was the end of August. The train was already hot from the summer sun, but became more heated from the songs of the students, that seemed to arise from the blood of their youth. The students removed their jackets, unbuttoned their shirts, and their sunburnt faces shone with sweat. An animal smell flowed with the heat, and when we finally fell asleep, tired from singing, we were already in Tokyo. There is probably no need to tell about the games. Under the unmerciful August sun, the red and white flags waved and danced, drums beat, and the cheers and the songs rang out across the field.

That year, the Third High lost again. Dead with fatigue, I

arrived home on August 31. For the newly entered students, it meant the end of their first summer vacation; the second semester was about to begin, and there might have been some who thought of studying seriously. Others probably thought that it would be another beautiful season of play. What did I think? I do not remember. Perhaps the excitement of the games had not wholly left me.

At home, I had to report on the battle of the two schools. There were stories that had to be told, even though I was dead tired. When I finally went to bed, it was past midnight. The next day, the first of September, was too hot to be called autumn. Sometime near noon, a great earthquake struck the Tokyo area.

The Great Eastern Earthquake of September 1 of the twelfth year of Taisho created its own mental shock in our household in Kyoto. Had we won the game against the First High, we would probably have stayed in Tokyo for an extra day. What would have happened if we were caught in the middle of the earthquake? Would I have followed my father's example and pursued geology? There is no point in speculating about this but in any case, I would probably not have followed my father's example.

That reminds me, however (although it moves back the story a bit), that I was encouraged to go into geology by my father when I was in the third year of middle school. I had not been much attracted toward geology, and indeed, it was one of my least favored subjects. In the Third High I learned geo-minerology from Shingo Ebara, who was a very enthusiastic teacher. A few days before the examination, numerous mineral samples were placed all over the room. Students had to walk around the room and memorize the names. During the exam, several of the samples were given to the students to identify. I was not very good at this, and could supply only some half-remembered names that were usually wrong.

My father, on the other hand, was gifted with powers of

observation and memorization of natural objects and natural phenomena. He used those powers as a foundation to raise his imagination. Geology, geography, and archeology were well-suited to him. I knew my powers of observation and memorization were not sharp, and I trusted instead my own power of logical thought. I had to choose a field of study in which I could use this power as the base from which my imagination could take flight. Later it became apparent that I had no choice but to do what I had done.

However, when I was a student, I was a "confusion" without a clear outline. One day, my father appeared before me with a large book in English, a college level geology textbook with many photographs and diagrams. "Read this book," he said. "If you find it interesting, then you should go into geology." The eldest son chose metallurgy; the second chose oriental history. My younger brothers did not seem interested in natural sciences, and my father probably thought it was not a bad idea for at least one son to follow in his footsteps.

I began to read the thousand-page volume, as my father had instructed me. At about that time, a questionnaire about college was distributed at school, and I wrote "geology" on the line requesting "desired major subject" without even thinking about it. But the books my father kept bringing me began to weigh heavily on me; I began to doubt whether I would go through them all. His study contained so many of these books, and I was already sick of them within a week. My interests became more strongly focused on physics than before. My father noticed that I did not read the books he gave me, but he did not say anything. I was sorry in a way, but I had already made up my mind. When a second questionnaire was distributed, I wrote "physics," without hesitation.

It is not easy to tell where the path of life diverges or divides. Even if I had been present during the great earthquake, I would not have chosen geology as my path. However, the reason I did not choose mathematics was clearly due to a specific incident;

but this, too, might have been influenced by my mind that had unconsciously been waiting for an excuse to leave mathematics and take up physics.

Although I was distracted by the Anniversary Festival and the cheering squad, I made frequent visits to the school library. My introversion nurtured my reading habit. I looked for difficult books, but they were not necessarily matched by my development as a human being. My extreme introversion had turned my eyes away from the real society. That I had shown no interest in the student strike, although I had been to the front gate at the time of the strike at Third High, was only one example of my attitude.

Although I was a member of the cheering squad, I did not think at all about the meaning of the cheering. I did not enjoy it altogether, but neither did I turn a critical eye toward the squad. As far as those things were concerned, I was only a child. In my case, the growth of understanding of mathematics, physics, literature, and philosophy did not coincide with my understanding of the real society. There was a large gap, a lack of harmony, an imbalance.

When I compare my own youth with the young people of today, I am vastly surprised by the difference. In a word, today's youth seems incredibly precocious. It is, perhaps, natural for today's youths, who are born into a more open and more stimulating society, to mature more rapidly. Nevertheless, they seem to possess an imbalance exactly opposite from my own. That is not a phenomenon that is peculiar to Japan, for all modern industrial societies seem to possess similar problems. Do these problems occur independently, arising from similar backgrounds, or do they start in one place and become transmitted to other areas of the world by mass communication?

The other day, I spoke with a young friend of mine on this very subject. He said: "Today's young people probably think that your youth was a very dull one." I asked: "Then are they, today's high school teens, satisfied with their lives?'' ''Of course not,'' he answered. ''They feel an emptiness. That is why they do the things that are reported in the newspapers. But *your* youth, it's too —,'' and I interrupted: ''Too childish, even to think about?''

Chapter 9

The Narrow Gate

Can a human being grow, at the same time maintaining a perfect harmony? Is it not the case that any period of growth, viewed in retrospect, shows great imbalance? After the Second World War, when the Japanese economy was in its darkest days, many who were very young had to endure the buffeting of the rough waves of society. It was natural for them to develop an interest in the nature of society. By comparison, the time of my boyhood was, at least for me, quite peaceful. My parents provided for my education, and there was no necessity for part time work by students as there is today. Of course, there were some prototypes of the behavior of modern high school teenagers. The parents of conservative families called them delinquent. One often heard descriptions like "delinquent youths," but it did not really sound too serious. There was that sort of "delinquent" even in the Third High School.

On the other hand, there were some who turned a highly critical eye on the society in which they found themselves. They were older than I, both chronologically and temperamentally. They must have known many things that I did not, but I did not particularly envy them.

It was certainly unbalanced for me to have poured almost all of my energy into reading, and to have all my thoughts centered around my reading. This tendency, to a degree, remains with me, and I do not admire that in a human being. But, suppose this imbalance had not existed in me; did it not have a lot to do with my becoming an established researcher of physics fairly quickly? I cannot believe that I had a well-rounded character at the stage of transition from boyhood to youth. Perhaps that was fortunate for me; although I was ignorant of many things around me, I insisted upon reading difficult books.

The first books I read with interest at the Third High library were on philosophy. My interests moved from Chuang-tzu and Lao-tzu to European philosophers. Kant was dominant in that period, but Bergson philosophy was also popular. Like many other Japanese youths of that time, I was drawn to Nishida philosophy; however, a curiosity about twentieth century physics began to divert me from my philosophical interests. Hajime Tanabe's *Outline of Science* and his *Recent Natural Science* attracted me more than any purely philosophical work.

My interest in mathematics waned slightly after middle school because the mathematics we were taught became somewhat more dependent on memorization. In algebra, there are many equations that must be memorized before moving on to the next topic. Solid geometry was Euclidean, so its theory was clear, but there I had a problem with the instructor.

The teacher of solid geometry gave lectures that were well-organized, I admit that. But he became irate if all the students were not meticulously taking notes, and he lectured very rapidly; to keep up with him required hard concentration. Soon after entering the Third High, during this teacher's lecturing, on one occasion I let my hand drop, because I did not hear a part of the lecture. Perhaps that was because I was not used to taking notes.

When he saw that my hand was still, his eyes became resentful, and he asked in a hard voice: "Mr. Ogawa, what *are* you doing?'' The pens of the class stopped writing; startled eyes were focused on me. I looked down, gripped my pen once more, and poised it above the paper. Then the teacher, ignoring — no, satisfied with the shock he had inflicted on the students — continued lecturing at an even faster pace. I left two or three lines blank and wrote desperately. At that time, I had not even a moment to feel resentful toward the teacher's unreasonable strictness. However, after class, I could not understand why I should have been so harshly confronted. That was not mathematics; it was more like army training.

One of my classmates was found to be ill with a weak heart, and he could not take notes at such a high speed. Perhaps he had been warned against exertion by his doctor; perhaps he was attending classes, although he should have remained at home. It is not always necessary to take notes in order to understand a lecture; some students can learn the important features on the spot. In any case, he did not take notes in any of his classes, but listened attentively.

However, when it came to the solid geometry class, the teacher acted as though he had been humiliated by this student, and this time he did not finish his scolding with a few words. The student stood there calmly and tried to explain his circumstances, probably thinking that any person would understand. But the teacher was unreasonably angry and would not heed his words. ''You have no need to make excuses,'' he said, ''for you will not have to attend my lectures after today.''

The student went pale, realizing the importance of the matter. Not having to attend the lectures meant that he had failed the course. Had it been intended as a joke, even a new student would have had the wit to answer: ''Yes, sir!'' with a smile, but this situation was bad. When the student went to the teacher's home to appeal to his judgement, he was turned back at the gate. His classmates sympathized with him, and they were outraged by the teacher's behavior.

When the First High-Third High competitions were over, the second term began immediately. Although the mountains surrounding Kyoto still wore their summer hues, the students, as they gathered, passing under the old gate, with "Hi! Hi!" to each other, seemed to bring an autumnal freshness; their very words seemed to brim with youth. Some of them had not left Kyoto during the entire summer, while some were returning from their hometowns with well-tanned faces. There were many topics to discuss; the students chatted cheerfully about their vacations and they were filled with various hopes. But when the bell rang for the start of classes, a current of uncertainty flowed among them.

On the first day of the new term, each teacher read the names of those who failed or got "warning points" for the previous term. Less than sixty points was a failing total; between sixty and seventy were considered "warning points." Some expected to fail, even before their names were read, and many more were wondering whether their scores fell above or below the critical line. In the solid geometry class, the teacher announced: "These obtained warning points," and looked over the room as he read the names rapidly. When I heard him read "Ogawa," I could hardly credit my ears, for I thought that I had done well in the examination. I could not believe it, and was sure there was some mistake. To this day I can recall my surprise!

I was doubtful that I had heard correctly until the teacher returned the examination papers. On my paper, the third of the three questions was indeed marked wrong; thus, my total points was sixty-six. Rapidly examining my work, I saw that the proof I had given was correct. Why was it marked wrong? I consulted my friends, and they agreed I was right, but one of them said: "You must prove it in the same way as in the lecture, otherwise this teacher will mark it wrong."

Thus, I could not protest. I had not remembered the teacher's proof, and had proved the theorem in a different manner. However, I was relieved to find that my proof was not wrong, and no longer cared about the points; but I could not overcome the negative feeling I had developed toward mathematics. What had so readily moved me away from mathematics was the teacher's method of awarding points. The boy, in his indignation, decided that he would never become a mathematician. A subject in which one must always answer in the way that the teacher taught — he would not dedicate his life to that!

The world of the mathematician disappeared from my view. That I happened to encounter this particular teacher might have been just a stroke of fate. In any event, I admitted to myself the importance of mathematics in becoming a scientist and later, when I studied the differential and integral calculus, the so-called higher mathematics, my interest revived to some degree. Actually, my eldest brother, Yoshiki, taught me the elements of the calculus when I was still in middle school so I did not find higher mathematics very difficult. Much later, I discovered it to be an interesting subject, and I learned that there was the joy of creation in mathematics. Still, I am glad that I did not become a mathematician. I find my excitement in the leap of imagination, and do not take much pleasure in the unfolding of an argument with perfect reasoning. Furthermore, it suited my temperament to be tormented, as physicists are, by the differences between the ideal and the real. There were also cards in my hand, other than mathematics, that had to be discarded sooner or later. I had to give up engineering, one of the reasons being that I was bad at drafting, although there were also other, more compelling, temperamental reasons.

Masao Fukuda taught drafting at the Third High, and although he was usually polite, he could be very sarcastic. A student in the drafting class, Seigo Yamamoto, was a good friend of mine. The teacher had stepped into the hallway for a moment while the students were busy with their drawings in the large classroom, twice as large as any of the others and filled with broad desks. Suddenly, a loud song welled up from the front of the room: "Oh, Susannah, don't you cry for me ..." The voice echoed through the room; it was Yamamoto, singing while he worked. It created a stir in the quiet room, and some laughter. "Be quiet!" someone said. Another said: "Keep going, Yamamoto!" When he had finished the first verse, from far away we heard: "You're pretty good, Yamamoto." It was Fukuda-san, the teacher, standing in the doorway and grinning.

I sometimes went downtown with my friend, Yamamoto. Once we went to a theater in Shinkyogoku to see Cecil B. de Mille's film, "The Ten Commandments." It was the first one I had seen since my grandfather took me to see the samurai films, so it made a big impression on me. However, I felt guilty about having gone to Shinkyogoku.

The drafting teacher, Fukuda-san, had a moustache, and his face was not stern; still, he made me uncomfortable and I found the hour of drawing to be practically endless. We were to draw lines with special pens on a thick paper called Wattman paper. I had no self-confidence, and was continually stirring the ink and sharpening the pen — anything to kill time. If the pen were too sharp, the paper would be cut as though by a knife. If it were not sharp enough, then the ink smudged the line. Once I felt I had to leave the classroom, because the hour seemed so long. When I returned, after a walk about the school grounds, I found that the ink in my pen had dried and the teacher was standing beside my desk; I broke out in cold sweat.

Things became quite hectic when the day to submit a drawing approached. I scrubbed at the mistakes with the ink eraser. I redrew it, and made another mistake. I erased again — again a mistake. Finally, the thick Wattman paper almost developed a hole. I finished with it at last and presented it gingerly to the teacher. He lifted it up against the window light and said: "Well! I can see Mt. Hiei through it." The only reason I did not fail the drafting course was that I passed the written exam at the end of the term.

Students who planned to study engineering, or especially architecture, at the university, were very good at drafting. I could not understand why their drawing ability was so much superior to mine. From childhood, I have never been good when beginning things. I feel bad when I realize my own gracelessness, and much worse if I think that a lot of people are watching me. The reason why I could never do the simplest exercises on the horizontal bar was probably because I could not do them at the beginning, when everyone was watching. My problems with drafting were not caused only by clumsiness.

However, after I have got over the first hurdle, I am able to move forward with determination. Of course, the road does not always widen as one goes forward; sometimes it narrows, or climbs a hill. Still, once I have passed a certain point, I never turn back. Whether I am able to jump the first hurdle depends largely on my own preferences. When I was small, I disliked some food, and could not understand why adults considered certain fish, like red snapper and bonito, to be delicious. I liked dried sardines and salted salmon much better. I liked horse beans and soy beans, but disliked most other vegetables. I did not eat apples or tomatoes until much later. I began to regret my stubborn preferences, my narrow-mindedness, after leaving elementary school. Still, at the bottom of my heart, something always made me choose one thing as my favorite — whatever it was, it was beyond my conscious control.

During my second year, the physics course began, and I found it much more interesting than middle school physics. For a long time there was a well-known physics teacher at the Third High, Sōnosuke Mori, but he was abroad when I began to study physics there. (It was a very rare thing for high school teachers to go to foreign countries.) We learned our physics from Ichinohe-san and Yoshikawa-san. The latter used an American textbook by a man named Duff. It contained sample problems at the end of each chapter, and I began to solve them as fast as I could.

I had not yet decided to devote my life to physics. Although the paths left for me to choose from were becoming fewer, there still remained more than one. I was so enthusiastic about solving the problems, not only because of my interest in physics, but because I was anxious to test my abilities. After I had solved one problem, I went on immediately to the next. It filled me with that mysterious joy that is granted to those who accomplish difficult tasks.

As far as physics was concerned, I never had to study specially for the tests. Many students tried to solve those problems just before the exams, but they did not have enough time. About a week before the scheduled exam, my classmates came, one by one, to ask me for the solutions to the problems. It became very time-consuming to explain the same problem to each of them, and I was annoyed. Then someone had the notion of gathering all the students together in a classroom, and I was to give a lecture. It was getting ridiculous!

Being naturally introverted, I was rather disconcerted, at least on the surface; but I also felt a certain pride. I stood at the lectern before twenty to thirty classmates, opened the textbook on the teacher's desk, and looked around. Everyone looked very serious. I had just turned seventeen, but those listening were older. Some were in their twenties; one had a noticeable beard. So far as I was concerned, they were adults. Feeling a secret superiority, I began to write on the blackboard. However, I did not get a perfect score in the examination. The obvious reason was that I had not done any cramming, but it was also due to my poor memory.

I was fairly serious about the experimental work also, but when it took a long time, I wanted to finish up and go home. One afternoon we were doing an experiment on the measurement of the electrical resistance of a blue copper sulfide solution contained in a U-shaped tube. (To this day, whenever I see blue neon signs, I remember that beautiful color. Was the experiment so memorable? It was not particularly difficult; it was just that a foolish mistake made the blue of that copper sulfide quite unforgettable.)

My lab partner was Jiro Oishi. We gradually increased the concentration of the copper sulfide solution, and each time we did, we measured the electrical resistance. The process had to be repeated many times, and it was going to take a long time. "Could we not do something to speed this up?" I asked. But Oishi only cocked his head. Then I spotted another U-tube on the table. "If we use two U-tubes, we could finish more quickly."

We divided our work then, one measuring the resistance while the other filled the extra U-tube with the solution required for the next measurement. That saved a substantial amount of time, and the experiment moved forward rapidly. However, we discovered that we were getting very unexpected results. As we increased the concentration, the resistance should have decreased; but that was not our result! The resistance was alternately increasing and decreasing with successive readings. What was going on? We looked at each other in puzzlement, unable to understand. No matter how hard we thought about it, we could find no mistake in our measurement technique. Time passed mercilessly; the other students, who had been doing other experiments, had all gone home. It was dusk outside, and lights could be seen in the windows of faraway building. The wind was coming up, also; and our experiment was lost in a maze.

"Oh!" I said finally, and almost at the same time, Oishi said: "That's right!" I do not know who saw it first. "Do you know the answer?" "Yes." We finally recalled something about the electrical resistance, and compared our two U-tubes. One was definitely thicker than the other. Obviously then, the readings would alternate. "How ridiculous," said Oishi, but I was too discouraged to say anything and just smiled at him in resignation. I was angry with myself for not having noticed such an obvious thing.

Although I sometimes used to make this kind of mistake, I did not come to dislike the physics experiments. My interest in physics gradually deepened, and I became dissatisfied with the physics I learned at school. I began to frequent the large *Maruzen* Kyoto bookstore that was located on Sanjo Street. Its imported book department was divided by subject matter, and I always lingered longest in the mathematics and physics sections. One day I found a book entitled *Quantum Theory,* written by the German physicist, Fritz Reiche, translated into English, and I bought it. With my knowledge of only high school physics, it was hard to understand. In spite of this, or rather, because of this very difficulty, I found this book more interesting than any novel I had read.

In the year 1900, the German physicist, Max Planck, discovered a completely unexpected discontinuity in nature, something that entirely destroyed an idea that had been accepted by modern physics: the idea handed down from the ancient Greek philosophy, through Leibnitz, that "nature does not take flight." The blow dealt by Planck's quantum theory to the classical physics, that had been thought to be within sight of completion by the end of the nineteenth century, was a strong one. But to destroy the practically completed, grand, and beautiful structure, and to build a new structure in its place was an extremely difficult job.

Neither Planck himself, nor Niels Bohr, who in 1913 achieved great success in applying the quantum theory to the problem of atomic structure, wanted to bring down the whole of the building. Rather, they wished to retain as much as possible, and to remodel or refurbish, rather than to rebuild from the ground up. Today we refer to this as the period of the "Old Quantum Theory" or "Early Quantum Theory." The remodeling did not proceed smoothly, for as one hole was plugged, another would appear. Some of the younger physicists were not contented with internal modification, and the plan for a new structure had begun to take definite shape.

The thirteenth year of Taisho, when I was reading Reiche's book (that is to say, 1924) was the very year when this transition from the old quantum theory to the new form, today's quantum mechanics, was taking place. At just about this time, de Broglie, in France, was about to publish his theory of *matter waves.* Of course, Reiche's book contained none of the new developments, and I myself knew nothing of them. Still, I could feel that theoretical physics was in a state of confusion, with discrepancies to be seen everywhere.

Reiche's book closed with these words:

> Over all these problems there hovers at the present time a mysterious obscurity. In spite of the enormous empirical and theoretical material which lies before us, the flame of thought which shall illumine the obscurity is still wanting. Let us hope that the day is not far distant when the mighty labours of our generation will be brought to a successful conclusion.[1]

Never, in my life, have I received greater stimulation or greater encouragement from a single book than I did from that one. I no longer own this book, for about twenty years ago, I parted with it in order to buy new books. It is a pity, but at the bottom of my heart it remains with my happy memories of those days.

Although I parted with Reiche's *Quantum Theory,* my study still contains many foreign volumes that recall old memories. Through the glass door of my bookcase, I can see a five-volume set in German; it is Max Planck's *Theoretical Physics.* Volumes 1 and 3 have grey covers and are printed on cheap paper, as were many books published in Germany after the First World War. Volumes 2, 4 and 5 were published some time later; their paper is of better quality and the covers are red with gold lettering.

It is the cheaply printed first volume that is connected in my mind with an important memory. Soon after I had finished reading Reiche's *Quantum Theory,* and as I was walking on

[1]Fritz Reiche, *The Quantum Theory.* English translation by H. S. Hatfield and H. L. Brose. (New York: E. P. Dutton, 1923).

Teramachi Street and came to the corner of Marutamachi Street, I noticed a bookstore with a large sign that read, "Books in German." The high school students of those days had a special fondness for the German language; for example, it was fashionable to insert German words like, Mädchen and Onkel, into the conversation. There was also an admiration for the German science that seemed to dominate the world.

Entering the bookstore, my eyes immediately found Planck's *Theoretical Physics, Volume 1*. The subject was dynamics, at beginning university level, and it seemed that I could understand it. Since the author was Max Planck, whom I greatly admired, I found myself walking faster than usual on my way home. I began to read the book as soon as I entered my room, and found unexpectedly that I could read it quite easily. It was written so that one could grasp the basic ideas, and the applications were also lucidly described. As I read on, I began to like Planck even more than before, and felt an increased regard for the quantum theory as well.

Much later, I became somewhat discontented with Planck's thinking — so simple and straightforward that it did allow much leeway. However, that he always thought things through completely, until he had satisfied himself — that aroused my admiration. At the end of my teens and the beginning of my twenties, I felt a strong affinity toward his simple, direct thinking, and I was very happy to think that I had something in common with this great scientist.

Planck was a professor at the University of Berlin. When I went to Europe for the first time in 1939, I spent about two weeks during August in Berlin. I walked at times around the University and thought of Professor Planck, and I wanted to meet him when the summer break was over. But the outbreak of the war tore me away from Germany, and thus fate stole my only chance to meet Planck. Later, I had the opportunity to meet all the other principal scientists who had constructed twentieth century physics, but I thoroughly regret that I never met

Planck, the father of the quantum theory.

In my third year at the high school, I took mechanics under Dr. Tateo Hori, an outstanding research worker in spectroscopy, the field of study that examines the light spectra coming from excited atoms and molecules, with the aim of discovering their structure. Behind the development of early quantum theory lay the perfecting of spectrum analysis. As nuclear physics does today, spectroscopy then occupied the center stage. Dr. Hori's lectures had the liveliness that one expected from a front-line scientist. Most students in the *Otsu* section of the natural science curriculum did not take mechanics, but only the smaller number of students in the *Ko* section. Among that group were Sin-itiro Tomonaga, Masatada Tada, and Norio Kobori. When we did mechanics problems, I found that all of them were excellent students. Tomonaga, I knew immediately, was smarter than any other friend I had ever known.

From this time on, Tomonaga and I were to walk the same path. How encouraging and challenging it was to have such an outstanding companion! As my father had criticized me earlier, I had a tendency to be stubborn. I try to solidify my ideas, and sometimes I go too far before I am aware; sometimes I jump too far. Tomonaga seldom made such mistakes, being the type of person who is aware of limitations, and yet comes up with clever ideas. He was a most valuable companion to me.

Although I have written at length about my classes and my reading, I was actually quite relaxed in my later years at the Third High School. Sometimes I participated in the interclass sport competitions, playing baseball and rugby, and taking part in rowing. I was not outstanding in any of these activities, and I never joined any sports club. There was never a danger that I would be made a team member, but it was fun to play as an amateur. Our class once took the school rugby championship; I was dragged out to play as a forward and lined up for a scrum. It was very tiring to play that game.

As the day of graduation neared, I decided that I would study physics at the university. The entrance examination was no problem; unlike today, it was the entrance to high school that was the main difficulty and the university entrance exam was relatively easy. A few more students than usual were enrolled in physics in my year. Aside from Tomonaga and Tada, there were also Kohei Kojima and Kiichi Kimura. We took the entrance examination at Kyoto University, and I did well in mathematics. The physics exam had a question on "Laue spots," and I did not remember having learned about them. I later shuddered to think of someone going to do physics and not knowing of the famous diffraction experiments of von Laue.

Chapter 10

Crystal

On a recent day when the weather turned summery, I walked over to the main building of Kyoto University, which I had not seen for some time. I usually spend my days in the Research Institute for Fundamental Physics, on the North Campus, not far from the main building located just across Imadegawa Street. However, because I was busy, and also lazy about going out for a walk, I generally do not step out of the Institute during the day unless I absolutely must.

Having finished my business at the main building, and feeling unusually relaxed, I took the time to look about. Directly in front of me was the main gate, painted white; beyond that, across the street, was the building of the Education Faculty, formerly housing the old Third High School. I walked toward the main gate, admiring the big old trees, and turned at the gate to face the building I had just left. Above it stood the clock tower; the two-story building extended symmetrically to both sides, faced with decorative chocolate-colored brick.

We usually called it simply the "main building," but besides an auditorium, the president's office, and administrative services, it also contained the large classrooms of the Law Faculty and the Economics Faculty. It was built when I was still in high school, but it still looked fairly new. As I gazed at the clock tower, memories flooded back to me. When I matriculated at the university, its president was Torasaburo Araki. Araki-san's characteristic large head appeared in my mind's eye.

A bright whiteness in the shrubbery surrounding the building caught my attention, as though zinc white pigment had been thrown into it. In the early summer sun, the torisan tree was in full bloom. There were no students present; the afternoon was quiet. The atmosphere of some thirty years ago, when I roamed here as a student, seemed still present. My college life began here. To the left of the clock tower, an old two-story brick building came into view. Above its main entrance, facing east, a sign read "Engineering Faculty Fuel Chemistry." However, it was the former classroom building for the mathematics and physics classes of the Natural Science Department.

In April of the fifteenth year of Taisho, my university life began here, but the building was already here in the thirtieth year of Meiji (1897) when Kyoto Imperial University was built. Since the building was owned by the Third High School before it belonged to the University, it possessed a longer history than the University of Kyoto itself. Most of those who were doing research in this building, most of those whose lectures I attended, are not in this world today. Professor Matakichi Iwano, from whom I learned thermo-dynamics, and Professor Bosaburo Yoshida, from whom I learned electrodynamics, are no longer with us. Professor Tamaki, with whom I was closest, Kajuro Tamaki, who lectured on mechanics, is gone. Of all the physics professors of those days, only Professor Masakichi Kimura, who lectured on optics, is still well.

My three years as a student at Kyoto University were relatively uneventful. In high school I had been involved to some extent in sports and the cheering squad. From my new university viewpoint, those past activities seemed pointless. In high school, I had changed my aspirations several times, but when I entered the University, all the paths except that of physics research disappeared from my view. There appeared to be no other choice but to walk this one path diligently; at that point, the road had not yet become rough.

I have many memories of those years. At the time I enrolled in the University, my father was dean of the Science Faculty, a position he did not want. Because everyone had insisted, however, he accepted the position with the condition that he would serve for only one year instead of the usual two-year term. During the entrance ceremony, each student came before the dean of the faculty to which he was pledged and signed an enrolment list. My father sat behind his desk silently as I signed; I found it embarrassing, even though my father had begun to act more kindly toward me. The cause of the change might have been, at least in part, that he had heard from Professor Sono that I had scored very high in the mathematics entrance exam.

The Natural Science Department at Kyoto ran on the credit system; one had to accumulate a certain number of credits to graduate in three years. The courses required to obtain the necessary credits were not rigorously prescribed; instead, there was something called a "typical program," showing a minimum of requirements for each major subject. One did not have to follow the typical program exactly. That sort of flexible system was congenial to one who had experienced the liberal education of the First Middle School. and the Third High School. Thus I listened to many mathematics courses, without being concerned about the "typical program." One of the lecture rooms was at the very back of an old brick building and I often sat there, in the middle of the room, taking notes with concentration. I also took the mathematics discussion classes very seriously.

Kiyoshi Oka, a young lecturer, conducted the discussion section for differential and integral calculus. He was a classmate of my oldest brother, Yoshiki, and I had heard Oka-san spoken of as a great prodigy. That meant that he had a tremendous memory and the deductive powers of a genius. However, Oka-san did not give the impression of being a university teacher and looked more like a member of the Third High School cheering squad, with a dirty towel hanging from his belt. The first problem that he assigned seemed frighteningly difficult, far beyond the capability of the students. We just sat there perplexed, but at the same time I felt a kind of thrill at the challenge of that kind of problem.

However, my interests soon concentrated on the new physics, the new quantum theory. In 1926, physics was like a huge ship tossed about in a fearful storm, or else like a land experiencing an earthquake; the shock could be felt even by me, who had just entered the gate of physics. Soon after entering the University, there was a lecture, sponsored by the Society of Arts and Sciences, on "The Present and Past of Physics." The lecturer was Hantaro Nagaoka of the University of Tokyo.

For a long time I had known the name of Professor Nagaoka and that he was the most important scientist of Japan, although I did not know why he was important. In the large lecture hall of the building with the clock tower, I listened to this great scientist speak to a large audience. The impression upon me was profound; I felt he was the greatest person I had ever heard. His message that physics, for the twenty years since the appearance of quantum physics, had been undergoing a far-reaching metamorphosis, moved me. Although he was about sixty at the time, I was struck by his youthful passion, like that of a student, combined with vast knowledge. Vaguely I realized that something new was going to appear beyond the old quantum theory that I had learned about from Fritz Reiche's book.

Soon afterwards, I bought a book that had recently been published in German, entitled *Mechanics of the Atom,* by Max Born. It was a thin book, no more than 110 pages, but the material in it was all new. Born was a professor at Göttingen University and the teacher of many brilliant young theoretical physicists. The year before I entered the University, one of his students, Werner Heisenberg, had published a new quantum theory at the age of twenty-three. Born immediately recognized its value and he worked together with Heisenberg and another young student, Pascual Jordan, to perfect the theory.

Born's book skilfully explained the just perfected — no, the rapidly developing theory. As for me, the new quantum theory, though attractive, was very difficult. From that time on, Max Born became one of the scientists I most admired. At the end of 1949, returning to New York from Stockholm, I stopped over in Edinburgh, Scotland, to visit Dr. Born, who had been chased out of Germany some years earlier and had settled down at Edinburgh University. The old gentleman who came to greet us when we arrived at our hotel in front of the train station, was Max Born himself, exactly as I had imagined him.

There are different types of scholars, and they can be classi-fied as "hard" or "soft"; Max Born clearly belonged to the type with more softness. I think that I, myself, am also one of the soft ones, and perhaps I was unconsciously looking for a similar temperament among the great scientists who were my seniors.

I became a devoted follower of the new physics during my first year at the University. It was far more meaningful than cheering at the high school athletic competitions. Then, I had had neither the ability nor the desire to become a team member myself, but this time it was a different story. I might be able to contribute something to the new physics myself.

When I was a university student, I usually stayed with my oldest sister Kayoko when I paid a visit to Tokyo, because my brother had obtained an appointment at Tohoku University after graduating from Tokyo University and had consequently moved to Sendai. My sister's husband was working at the electrical testing center of the Department of Transportation, and the family had moved to the Tokyo suburb of Omori. He was a mathematics fan, and as soon as he saw me he would start an argument: "Why do you like the new physics so much? Isn't mathematics better? You can prove whether you are right or wrong in mathematics.'' My brother-in-law liked European music and although he practised Yokyoku,[1] he was uninterested in other Japanese music. He liked those things that are pure and unambiguous.

I defended my own position vigorously: "It is because you cannot tell what will be the outcome of the new physics — that makes it interesting." We argued endlessly.

My sister, who had mostly literary tastes, came and listened to us, but when the topic turned to mathematics versus physics, she would smile and say, "There it goes again!" The couple's eldest son, Iwao, was small at the time, so he probably did not understand our arguments. Iwao became a physicist thirty years later, but that does not mean that I won the argument, for boys tend to ignore their father's advice; I was that way, too.

Koshiro Takéi, the husband of my second sister Taéko, never tried to argue with me. Specializing in city planning, he had ambitious dreams and great enthusiasm, which may be what induced my father to allow his second daughter to marry him. About the time of my entering the university, Taéko returned to Kyoto with her husband and her many children. Koshiro Takéi had become a professor of engineering at Kyoto University; our house became lively once again.

We had moved again several times. Once we were in Shimogamo, but quickly moved to Tōnodan, and then we moved once more to a house that was four or five houses away. That was the white-walled house of Bishamon-cho, Tōnodan. For the first time in his life my father bought a house rather than rented one, and this house became his last one.

The name Tōnodan[2] seems to have originated from a tower of the Shokokuji Temple that was there long ago. The tower was

[1] The sing-song narration of the Nō plays
[2] Meaning "steps of the tower"

burned to the ground in an ancient battle, but its stone steps remained. The area had belonged to the temple, and for a long time it was covered by a bamboo forest. About the thirtieth or fortieth year of Meiji, the bamboo groves of Tōnodan were developed into a residential area; several of the first professors of Kyoto University built houses there. In particular, two large houses with earthen walls were built, one at the northern edge and one at the southern edge. The northern one belonged to Byonnojo Mizuno and the southern one to Hanichi Muraoka, both professors of physics.

Long before my entering the University, Professor Muraoka had retired and moved his home to Isé. The house in Tōnodan had changed hands and was bought by my father. To the south and west sides of the thirteen hundred square meter lot, there were narrow roads. The white exterior wall ran the length of those roads; passers-by called it "The Great Wall of China." It had been possible for a university professor of the generation earlier than my father's to build a new house requiring so large a lot. That was not the case during my father's time; it was quite difficult even to acquire the old one. There were big outlays required to educate the many children, and because my father had such wide interests, he continued to buy expensive old books, antiques, and swords. My mother frowned at each of his purchases.

Just before buying the house, Mother was saying, "At this rate, we will live all our lives in rented houses and will end up with nothing. It is now a little difficult, but I would like us to buy a house, even if it means going into debt. That way, we will at least own the house." I did now understand very well the household finances and as I listened, I began to realize that I could not depend on my parents forever. But what should I do? I had nothing planned, and in any case, I could not alter my strong desire to study physics. I believe that we bought the house a little while before my father retired from the University, but I am not certain. In any event, I certainly commuted to Kyoto University from Tōnodan when I was a student. In my first year, mathematics occupied much of my time. In physics, I began with Professor Tamaki's "Mechanics" and Professor Ishino's "Thermotics."

Professor Kajuro Tamaki wrote in elegant alphabetical script on the blackboard. The vector symbols seemed to flow; the writing was like a Japanese poem written in beautiful cursive calligraphy. He was always well dressed, and even his small gestures gave the impression of his being a British-style gentleman. He had, indeed, studied at Cambridge University. Accomplished in the fields of hydrodynamics and relativity theory, he loved the finished beauty of classical mechanics as well as the relativity theory that was its extension. The quantum theory, still rough-hewn, seemed not to agree with the elegant professor's taste.

Professor Matakichi Ishino preferred to use German, and even his features gave the impression of *Herr Professor*. The main part of his thermotics course was thermodynamics, the field that Max Planck, whom I greatly admired, brought to perfection. After laboring to derive the characteristics of heat radiation from thermodynamics, Planck had given birth to the quantum theory. Thus the thermotics lectures seemed to be closer, in many ways, to the new physics than did mechanics. In the third year, I would receive private instruction from one of the physics professors. Sometimes when I thought in an indistinct way about the two years ahead, I told myself that either of the professors, Tamaki or Ishino, would be an attractive choice.

Generally speaking, my college career began well. Since I had entered Kyoto University from The Third High School, I had no reason to be disoriented. My classmates in the Physics Department were fewer than twenty in number and included four who had been my friends since middle school or high school: Sin-itiro Tomonaga, Masatada Tada, Kiichi Kimura, and Kohei Kojima. Among the mathematics majors, Norio Kobori and Yoshiro Mori were my old acquaintances. All these people later became professors, and they still walk the path of science.

In the university there were no longer gymnastic classes or drafting classes; there were no sports or cheering squads. I could study whatever I liked. To some extent, the pessimism of my middle school years remained with me during high school. I often felt depressed and suspected that I might have a minor neurosis. In college, that feeling disappeared as if it had peeled off like a piece of paper. The only problem I had was with glass-blowing (I shall speak of that later), but it was not that important.

There was, however, a small uncertainty in my future that concerned the medical examination for induction into the army, which every man was required to undergo. Near the end of my first year at the University, that is, in January 1907, I had my twentieth birthday, and in April 1 was to be examined. If I were classified A-1, I would have to serve in the army as soon as I graduated from the University. I became uneasy, thinking that my study might be interrupted at a critical point.

I was called to the place for examination on a day when the sun was already hot. I cannot recall whether it was the City Hall at Kamikyo or one of the school buildings. I was given the eye examination; and the rest was also quite simple, and I was moved forward quickly, passing the others. In front of me, another young man was also moving forward quickly and he seemed familiar. It was Takujiro Ishino, the second son of Professor Matakichi Ishino. He was one year behind me in school, and enrolled in the Medical Faculty of Kyoto University. As we were exchanging greetings, we found ourselves together in front of the officer-in-charge, who looked through our papers and said "Class C." Then, relaxing a bit, he said: "You are young university students. We cannot use you as soldiers, but you should study diligently, and in that way let the world know about Japan."

That was during a period of military reduction, when it was

not necessary to induct university students. My oldest and second oldest brothers were classified *B*-1 and *B*-2, respectively. Since entering college, I had not participated in any sports, and I spent a lot of time studying at home or in the library; I was thinner than in my high school days and my face was paler. I had expected to be classified B-2, but being *C* was a trifle disturbing. I wondered if it was because of my eyes, which were astigmatic as well as myopic. In any case, because of the officer's judgement, the problem of military service was removed from my concerns, and the days when I could study as I liked were going to continue.

There is a Chinese saying that "heaven does not allow man to rest." I often recall this recently, as so many duties fall upon me daily. Why am I always dragged into things that I do not want to do? Why do these inescapable duties bind me tenfold and twentyfold. Perhaps, in those days, heaven was allowing me more rest, or at least it seems that way. However, in a deeper sense, I was not permitted to rest even then, for the world of physics was changing urgently and constantly.

In 1926, my first year at the University, Erwin Schrödinger introduced wave mechanics and created a sensation in physics. There had been no such disturbance when, two years earlier, Louis de Broglie put forward the theory of matter waves, a forerunner of Schrödinger's theory; but this time there was a huge response, for several reasons. Many scientists had been taken aback by the complexity of Heisenberg theory and had come to dislike it, without really understanding it. Observing that Schrödinger's theory was much easier to understand, their interest was aroused. Another reason was the strong persuasive quality of Schrödinger's papers. The physicists were excited and Japan, too, began to feel the waves. I guessed what was happening from conversations between the professors and my seniors and I felt that I could not afford to be idle.

In my second year, I spent all my spare time in the physics library. I had no use for the old books that filled the shelves but wanted to learn, as quickly as possible, the articles that concerned the new quantum theory, those that had been published in foreign, especially German, journals within the past two or three years. As I searched them out, I began to feel that this was somewhat too ambitious for a university student in his second year. The papers that already were published were considerable in number, and new issues of the journals kept arriving and accumulating on the library shelves. I was bewildered, not knowing where to start, and for a time I nibbled at a wide variety of articles. Soon I decided to read Schrödinger's own papers systematically, as they were the easiest to comprehend.

On a shelf in my study, I can see Schrödinger's *Collected Papers on Wave Mechanics* next to the five volumes of Planck's *Theoretical Physics* and Born's *Mechanics of the Atom*. Black letters spell out the title in German on the red-orange cover. The title page is faded; the book is not soiled, as I do not have oily hands, but it shows signs of age. I was in the midst of reading those papers in various journals when I found a book containing all of them at the Maruzen Bookstore. During my second university year and at the beginning of the third, I was completely immersed in Schrödinger. His arguments were strong and sharp enough to convince any reader of his claims. Like Planck, his theories were simple and solid; he was trying to adhere to a "unified wave theory."

The book of Max Born that I had read a year earlier, had emphasized the discontinuity of nature. Perhaps Born had expected to be able to find an element of discontinuity in everything, including space-time, when he wrote his book. A year ago, I had wanted to follow this direction. Schrödinger pointed out an exactly opposite direction, emphasizing the continuity of nature, and he tried to adhere to that concept, that of a unified wave theory. I became attracted to this theory; the pendulum had swung within a year from one side to the other. Gradually, I began to realize that I had, in both cases, gone too far. Of course, much before my own realization, the world of physics had begun to construct a theoretical system incorporating both the continuous and the discontinuous sides of nature, uniting the views of Heisenberg and Schrödinger. There will be no end if I continue a discussion like this one, so let us return to the original topic.

My appetite for learning was strong, and there was a mountain of new knowledge to digest; it was like sitting down before enough food for several meals. I was only in the second year of college; I had to listen to lectures and attend discussions; I had to perform experiments. In the second year Professor Masamichi Kimura's optics lectures began. His specialty was spectroscopy, which was closely related to the new quantum mechanics, i.e., research on atomic and molecular spectra. I was interested in those experiments, and I teamed with Kiichi Kimura, who was more of an experimentalist than I. When he wanted us to continue working during the summer break, I made no objections. There were many employees who were working during the summer, but no other students. We made sparks between carbon and metal electrodes in the empty student laboratory, and we took photographs of the spark spectra. We went into the darkroom and cut the photographic plates with glass-cutters. Now that I have been away from the laboratory for so many years, the memories of this period are especially dear to me.

In the third year, I would have to decide on a specialty and to be under the direction of one professor, but although the end of my second year was in sight, I was still undecided. Among possible specializations, Professor Kimura's spectroscopy was the closest to what I desired. In March 1927, Kimura invited a young German physicist, Otto Laporte, to lecture on the theory of atomic spectroscopy. It was the first lecture I had heard in a foreign language, but I understood well both the words and their meaning. Perhaps it was easier, just because it was the English of a German! I understood the content because I had been studying optics. That was all fine, but the difficulty was that Kimura's laboratory did not accept students of theoretical physics.

If I were to do experiments in spectroscopy, I had to learn glassblowing. However, my biggest anxieties were aroused by the horizontal bar, drafting, and glassblowing. Just after beginning at the University, our class was told to make a simple device to measure an increase in volume, an instrument usually called a dilatometer. One end of a glass tube is melted in a gas flame and sealed off; the other end is stretched under heating into a very fine tube. It should have been quite easy, but I could not do it. I pulled the tubing out in the flame, but it broke before it was thin enough. My classmates had no difficulty in constructing their dilatometers. Even Tomonaga, who was no better at drafting than I was, did well at glassblowing. But I could not get the hang of it, finally gave up, and asked to be assigned a different experiment. If I were to be involved in spectroscopy experiments, I had to be able to bend and connect glass tubing as needed. Thus, I thought myself unqualified for the Kimura laboratory.

My father advised me to study under Professor Ishino and to pursue both theory and experiment, but I thought that doing both would be too heavy a burden for me. After several days of indecision, I visited Kawada-san, who was doing x-ray experiments in the Ishino laboratory. I listened to his explanation about what he was doing and found it interesting. At this point a stranger entered the room; one could tell that he was not a physics researcher. His conversation with Kawada-san was about business, about the type of machines to be ordered and their price. The conversation sounded like one from an entirely unknown world, and I realized that to do experiments I would have to deal with that sort of business. As I listened in silence, I thought that maybe I could only do theory, after all.

A bit later, I visited Professor Tamaki's research room with Tomonaga and Tada, and the professor accepted us warmly as his students. It was at this point that my path became that of theoretical physics alone. The Tamaki research room of that period was where the ambitious gathered. People who were years out of college were studying what they wished, about ten members altogether, a much larger group than in the other research rooms. Not surprisingly, several were studying hydrodynamics, the professor's specialty, and one was working on acoustics. The professor himself was very interested in music, and the research room contained organs and kotos. I had heard that Professor Tamaki played the koto, but I never had the pleasure of listening to him. He was also interested in the Japanese bells, and there was a small bell there — more like an alarm bell than a simple musical one. Sometimes one heard the sound of that bell coming from the research room.

Naturally, there were people studying relativity theory, on which Tamaki had published many papers in his youth. However, the presence of some who studied the new quantum theory was unusual; they were Sotohiko Nishida and Shohei Tamura. Professor Tamaki had little interest in the new quantum theory, and probably he was perplexed by it; but he always respected the freedom of the people in his research room. As long as one did not step beyond the boundary of theoretical physics, there was no pressure, no matter what one studied. Even if after several years one's work did not bear fruit, that person was not dismissed. Everyone studied at his own pace.

The atmosphere was different from that of the other research rooms, but being used to the policy of freedom of Sotosaburo Mori, I did not perceive it as strange. Rather, I finally chose this research room because I felt an affinity toward that kind of atmosphere. Of the three newcomers, Tada chose to study hydrodynamics, and Tomonaga and I chose the new quantum theory. We had a mountain of things to learn in both physics and mathematics. We had to study higher algebra, especially group theory — something that had been useless in physics up to that point. We had to study, at the same time, both classical physics and the new quantum physics. Professor Tamaki taught us wave diffraction, and Nishida-san and Tamura-san helped us with the new quantum physics, but actually we were studying almost on our own. I wanted desperately to catch up with the forefront of theoretical physics during my third year, which made it a busy year.

This is a bit out of the story line, but Sotohiko Nishida was the oldest son of Professor Ikutaro Nishida. I was a fan of the professor and thought it foolish not to hear his lectures while I was a student at Kyoto University, the home of this great philosopher. I do not remember whether it was during my third year or the year after graduation that I went every week to hear his lectures on philosophy. At that time, the professor's popularity among the young people was considerable. Some of the Third High School students came to listen to his lectures; the large classroom of the Law Faculty was always full. Each lecture was complete, like one story or a serial. He would walk to the lecture platform carrying five or six thick books, which he left on the table. Ignoring them at first, he walked from side to side upon the platform, while lecturing.

Professor Nishida was very near-sighted; his thick glasses sometimes gave off reflected flashes of light as he walked back and forth. He seemed to be talking to himself, rather than systematically teaching. From time to time, he would stop and open one of the thick volumes on his desk, the work of some famous philosopher. At the next moment, he would deliver harsh criticism against the author. I have forgotten the content of his lectures, but my powerful impression of this man remains. Much later, I sometimes visited the professor's home in Kyoto or in Kamakura. In ancient times, philosophy and theoretical physics were one and the same; today they are quite far apart, but when I was conversing with Professor Nishida, they seemed to come close together again.

In my house there is a calligraphic work by the professor that reads "Ho Ho Sei Fu." Whenever I see it, I remember with nostalgia the professor who walked about his Kamakura residence wrapped in a white obi, in a slightly bent posture, in deep thought.

In Kamakura,

here in a deep and quiet valley,

a man walks deep in thought.

Once I composed this poem.

As I was desperately trying to reach the front line, the new quantum physics kept moving forward at a great pace. The new theoretical system called quantum mechanics was beginning to be formed, and that irritated me. What lands would be left for me to pioneer in, if the world of the atom became completely understood by quantum mechanics? Was I too late in trying to became a theoretical physicist? Soon I began to realize that I had no reason to worry. It was true that quantum mechanics was near completion, that it was being applied to many areas, and was achieving remarkable successes. However, it could not cover everything. Of the two great pillars of twentieth-century physics, quantum theory and relativity, the former was highly developed, but had not yet been merged with the latter.

How was it possible to include the theory of relativity in quantum mechanics, to make relativistic quantum mechanics? I began to realize that this was the great homework assignment for theoretical physicist. However, in 1928, the young British genius, Dirac, discovered the relativistic wave equation of the electron. That was a great stimulation to me — or rather, it was something of a shock. In any case, I had to study Dirac's new electron theory. My graduation thesis, although it contained nothing that was creative, concerned Dirac's new theory.

My busy three years as a university student were about to end. I was still nothing, but the direction of my research was well established. Though no crystal of any visible size had appeared, the seed of the crystal was there. In the context of the closing words of Max Born's *Mechanics of the Atom:* "A single crystal is clear, yet a collection of crystal fragments is opaque." I was still a fragmented crystal.

All members in physics in front of Department of Physics, Kyoto University. Hideki is at the center in the back row. Sin-itiro Tomonaga is sixth from left in the second row from the back. Professor Kajuro Tamaki and visiting Professor Otto Laporte are the third and the fifth from left in the front row. 1928-1929.

Chapter 11

A Turning Point

In March of the fourth year of Showa, a little before my graduation from Kyoto University, I began to have misgivings. If I were to continue doing physics, it might come to nothing. I was so pessimistic that I even thought of becoming a priest. My distaste for society, present since middle school days, lifted its head at that time. A dislike for society resides within me even today, although it is more a desire to avoid contact than an active dislike. I want my interactions with other people cut down to about one-tenth, and I want to live quietly. If no one paid any attention to me, that would probably mean a lonely life; but to endure the loneliness would not be so bad, either. That sort of wish, a quite unrealistic kind of wishful dream, gives me comfort.

Perhaps it was not so unusual that, just before graduation, I wanted to get away from society. There was a temple called, Chokoji, near Osaka Castle, in the eastern district of Osaka. The wife of the priest of the temple was related by marriage to my father. The priest and his wife had no children, and my brothers and I appeared to have been prospective adoptees. We avoided Chokoji, saying, "We'll be made into priests if we go there." When I was depressed, I remembered Chokoji, thinking they would receive me gladly and make me a priest. I would think that way for four or five days; it was like an attack of mumps. After graduation, I entirely forgot those thoughts.

The Tamaki research room did not accept graduate students, so the three of us continued our research there as before, with the nominal title of (unpaid) assistants. That was a time of great economic depression, and jobs for college graduates were very scarce. Therefore, many of my classmates remained at the University; the depression had made scholars of us. My classmates, one after the other, let their hair grow long, although some already had their hair neatly parted at the graduation ceremony; my hair was still cropped short. Mother had a suit made for me, but I seldom wore it, going to the research room each day in my old student uniform.

About that time, half of the physics research rooms were supposed to change their location. East Oji Street, then called Toyama Street, had a tram that ran as far as Marutamachi, and the plan was to further extend the tracks to Imadegawa. That meant that the tram would run just to the west of the physics laboratories. The professors said that the effect of the trams on the galvanometers would make precise experimental measurements impossible, so that it was necessary to move the research rooms to a place that was more than one hundred meters from the tram line.

For that reason, a move to the North Campus was approved. But, if inability to do experiments were the only reason, then only the laboratories should have moved. The new modern building on the North Campus was not large enough to accommodate all of physics, and so only half of Tamaki's people were moved there. Luckily, those of us studying quantum physics received a room in the new building, transferring out of our old-fashioned one. I spent my days there, studying in a happy mood.

Looking back over my whole life in research, I view the three years after graduation as providing an extremely valuable foundation. The competitive swimmer glides under water for a brief interval after he has dived — it was that kind of preparatory period for me. I had two large research topics before me; not topics, really, so much as unexplored areas. The first was to develop relativistic quantum mechanics one step further. The second was to apply quantum mechanics to the problems of the atomic nucleus. Both of these subjects were much too ambitious for me, just out of the University.

Although I was only twenty-two, I was not lacking in years. At that time, most of the physicists who were contributing to the development of quantum mechanics were in their twenties, some only five or six years older than I. Four of the most outstanding scientists: Heisenberg, Dirac, Wolfgang Pauli, and Enrico Fermi were all born in the period 1900-1902, and they all had attained major achievements by about 23 or 24 years of age. In the autumn of 1929, Heisenberg and Dirac both visited Japan; attending their lectures was a great stimulus to me.

As to the two great problem areas that I had recognized, I was at a loss where to begin. At that time, research on the atomic nucleus was not in the mainstream of physics. A man ahead of his time, Ernest Rutherford, was achieving dramatic results on the physics of the nucleus, but the majority of scientists hesitated to enter that field. They were content to study the atomic electrons that moved around outside the nucleus. Why was it that most physicists avoided the nucleus? One of the main reasons was that the structure of the nucleus was incomprehensible. Many scientists believed that matter can finally be broken up into two or three elementary particles. At that time, the only objects recognized to be such particles were the electron and the proton — no, there is one other called the photon, but I will speak about that later. However, if all matter were made of electrons and protons, then the nucleus would have to remain a great mystery. That view made it almost impossible to understand the various properties of the nucleus. Recognizing this task to be impossible, most scientists stayed away from the nucleus. Some of them vaguely imagined that the electrons in the nucleus behaved in some very unusual manner.

So, this is what I was thinking: before considering the behavior of the electron inside the nucleus, one should study the interaction of an electron outside of the nucleus with the nucleus itself, in order to use that as a basis for further research. The method employed should be to study the hyperfine structure of the atomic spectrum.[1] Since Dirac's electron theory, especially, had achieved a remarkable degree of success outside the nucleus, one should apply this theory to the hyperfine structure of the hydrogen atom. That is how my research life began.

A hydrogen atom is made of an electron and a proton, bound together by their electrical attraction. Aside from this, there is also a magnetic force (though it is weak) because the proton and the electron are small magnets. No one had yet theoretically determined the hyperfine structure because of these magnetic forces. I tried to do that, and found that it was possible to draw several conclusions. Writing up these results, I gave the report to Professor Tamaki, who stored it in his safe, saying he would read it later.

Soon afterwards, a paper by Fermi on hyperfine structure appeared in one of the journals. I was very disappointed, for he had not only taken up the problem that I had started to work on, but he was also one step ahead of me. Having met with disappointment, just as I was about to embark on the problem of nuclear structure, my only thought was to change to another problem, for a while at least. It was just then that the great paper by Heisenberg and Pauli on quantum electrodynamics appeared. In certain respects, that article was the final balance sheet of the quantum theory originated by Planck.

When the quantum theory first appeared, it threw a cloud of mystery on the nature of light. By the end of the nineteenth

---

[1]The effect of nuclear magnetism on the atomic line spectrum

century, it was considered an unchallengeable truth that light was wavelike, a form of electromagnetic radiation. However, the quantum theory insisted that light must also have particle-like characteristics. The truth of the concept of light as a collection of photons also could not be denied. Thus, for at least twenty years, the wave-particle duality of light was the big question mark in the world of physics. After the appearance of de Broglie's theory matter waves, the duality puzzle was extended to particles of matter, such as electrons.

The mystery of the wave-particle duality was partially solved for matter by quantum mechanics. Similarly, to attain a solution of this puzzle for light, it became necessary to treat the electromagnetic field quantum mechanically. The quantum electrodynamics of Heisenberg and Pauli can be said, in that sense, to be a settling of accounts; it satisfies the requirements stated above. Nevertheless, there was an imbalance in this final audit. Infinity, which is a number that does not really exist, was written in one of the bottom lines of the audit! The balance sheet is one of energy: in the world of physics, built upon the principle of conservation of energy, the currency is energy, its unit not the dollar but the erg or joule. If the audit is truly balanced, the "number" infinity should not appear in the energy column. How could one eliminate infinity from the accounting of Heisenberg and Pauli? That was the new problem that their paper set before us. I read their report many times, and I considered thoughtfully every day how I could defeat that devil called infinity. But the demon was much stronger than I.

Our research room was located on the second floor of the new physics building, which stood on the grounds belonging to the Agriculture Faculty. Outside of the south window, one could see a grey building with a Nordic-style roof, having a steep slope. The wall of the building was ivy-covered; under it romped several mountain sheep, which bleated sometimes in a strange voice. As I fought daily with the devil called infinite energy, the call of those sheep sounded to me like the sneers of that devil.

Each day I would destroy the ideas that I had created that day. By the time I crossed the Kamo River on my way home in the evening, I was in a state of desperation. Even the mountains of Kyoto, which usually consoled me, were melancholy in the evening sun. Next morning, I would leave my house feeling renewed strength, but again would return at night thoroughly discouraged. Finally, I gave up that demon hunting and began to think that I should search for an easier problem. And while I was thus wasting my time, the area of application of quantum mechanics was expanding rapidly. From atoms and molecules to chemical bonding and the theory of crystals — quantum mechanics was succeeding everywhere. New scientific fields, such as the theory of the solid state and quantum chemistry, were about to emerge.

I had a great deal of free time to spend and decided to learn another language. I had no problems with English and German, having studied them in school. As for French, for a while I had attended an evening course during high school, but I was usually very sleepy by that time of day. As I listened to the lecture, my eyelids would fall of their own weight. No matter how hard I tried, I could not keep my eyes open. There were many such days, so that my French did not improve, and I could barely read French books and papers on physics.

I decided to attend the Japan-France School,[2] at that time located at Kujosan. Two or three days a week, I left the research room early in the afternoon and took the tram which used to run to Keagé, passing the zoo. I was fond of this tram, which was always nearly empty. The conductor collected tickets near the end of the line, and then turned the tram around by pulling on a rope. I enjoyed that relaxed atmosphere. The tram stop was near an old electric station, and the Miyako Hotel could be seen across the tracks of the train that goes to Otsu. I would walk along the tracks for a little while, then turn into a path that sloped upward to the left. Halfway up the hill was the Japan-

[2]Nichifutsu gakuen

France School, and from it there was a view over the entire city of Kyoto. That atmosphere, so different from the physics research rooms, was one that I really enjoyed.

It was the period when French films were beginning to be popular in Japan, the first one being "Under the Roof of Paris." I had stopped feeling guilty about going to see films in Shinkyogoku, so I went to see it. The theme song of the film, "The memories, ... ," was being sung all over Kyoto and I also learned the French words and sang, "When she turned twenty ..." The Japan-France School fitted the image I had then of France and Paris. The building, in the middle of a beautiful wood, the people who gathered there — everything contributed to form a milieu quite different from any I had known before. The people were young, and many spoke French very well. Their dress and manners were refined, especially those of the women, usually daughters and young wives of the prestigious houses of Kyoto and the Hanshin area.

I, on the other hand, still wore the university uniform and my hair was cropped short. I almost never spoke to the other people there, and seemed like a man from another world, accidentally appearing there. During the rest period between lectures, I looked silently down on the city of Kyoto. I had nothing else to do; but I was not unhappy to be there.

There was a French woman teacher, who seemed to be sensitive and kind. I was happiest during her class period. Once, being asked to write a composition for this teacher on "A Walk," I wrote these lines in French:

I do not wish to seek the strong stimulus of the city, and I am also too lazy to travel into the country, far from the city. Because the Imperial Palace is near my house, I often take a walk in its grounds. Autumn is the best time. The fallen leaves, which carpet the paths between the ancient trees of the Palace make tiny sounds under my sandals; these sounds remain in my heart as an unforgettable echo.

In one of the large open areas of the Palace, several groups of children have occupied the corners and are playing baseball. Under a large tree in the middle of the area, a boy on an errand pauses beside his bicycle to watch a game.

It is very quiet on weekdays, and often one can see young mothers pushing their baby carriages. On the lawn, two ginko trees are standing. In autumn, their yellow leaves cover the ground where they stand. One sunny morning, I saw two small boys shower each other with the leaves under the trees. A little dog came running, and all the three played upon the yellow carpet. Looking up above them, I saw one lonely branch of the ginko, bare of all its leaves, glowing a brilliant pink in the sun.

I always leave my house with the intention of thinking about a problem, but my attention is seized by the events around me at the Palace grounds, and forgetting everything else, I head for home in a state of pleasant relaxation ...

Indeed, I was a lonely stroller. Because of my quiet nature, it was not unusual for me to spend a whole day in the research room reading journals, without exchanging a single word with anyone. To my friends, I must have seemed unfriendly and uncongenial. I was not satisfied with my behavior, but found it hard to change. I decided that not only was I unhappy myself, but I was unable to bring happiness to others, and so I felt that I deserved to be lonely all my life. I had better not marry (so I thought); I shall simply lose my own freedom, and not be able to make my marriage partner happy.

When I was tired of studying, I sometimes drew a floor plan for a room to be occupied by myself alone. I still have one of those plans: there is a chair, a bookcase, and a bed in a room of area about fifteen square meters. It is amusing that storage space is provided only for books, but amazingly, there is also a bath. That described my fairy-tale world. No, it was too lacking in fantasy, too dry, and too close to the reality, to be called that. Still, the child who loved toy gardens had not left me.

The window of my small world opened out only to the garden of science, but from that window, enough light streamed in. In fact, Professor Masumichi Kimura brought in many scientists from the outside as guest lecturers, providing an important stimulus. During the two years following my graduation, Bunsaku Araktsu, Yoshikatsu Sugiura, and Yoshio Nishina lectured on quantum mechanics, presenting different points of view. All of them had studied the new physics in Europe; Professor Nishina had the greatest influence on us.

At that time, the phrase, "Copenhagen spirit," was frequently heard in the physics world, referring to the Institute of Theoretical Physics at Copenhagen University, with Niels Bohr as its head. The best theoretical physicists came from all over the world to learn from Bohr, including some Japanese scientists. Yoshio Nishina had a particularly long stay in Copenhagen. His lectures were not only explanations of quantum physics, for he carried with him the spirit of Copenhagen, the spirit of that leading group of theoretical physicists with Niels Bohr as its center.

If I were asked to describe the spirit of Copenhagen, I would not be able to do so in a few words. However, it is certain that it had much in common with the spirit of generosity. Having been liberally educated, I was especially attracted by that, but I was also attracted by Professor Nishina himself. I could talk to him easily, although I was usually very quiet. Perhaps I recognized in Nishina the kindly father figure that I could not find in my own father. Whatever it was, my solitary mind, my closed mind, began to open in the presence of Professor Nishina.

One clear day in the autumn of the sixth year of Showa, fate descended upon me in the person of Kosho Kishi, then the Secretary of Kyoto University, and afterwards, head of its administration. He brought with him an unexpected proposal of marriage. There may have been other marriage proposals discussed with my family, but they probably disappeared before I knew of their existence. In any case, I do not remember any other marriage proposals. The one made through Kishi-san as intermediary was the first one I was aware of — and it became also the last one.

The other party was the youngest daughter of the proprietor of the Yukawa Gastrointestinal Hospital. The Yukawa family originated, as did my own, from Kishu,[3] which gave my parents a sense of familiarity. My own main interest was, of course, in the personality of my prospective wife. Why did I, after secretly declaring my independence, become interested in a proposal of marriage? I cannot fully explain, but if forced to answer, I would say, "curiosity," at least at the beginning. Also, as a young man, I could not regard a marriage proposal as a blow to my vanity!

Certainly, a curiosity toward the other party in the proposal existed, however slight, from the beginning. Suddenly, however, that interest was greatly strengthened. That was because a photograph of her was placed before me, taken from a women's magazine. In it she was standing, wearing a kimono whose sleeves seemed to be weighted. Apparently, she was from a well-to-do family: she looked innocent; but her eyes were sparkling. Sumi Yukawa was about to greet the spring of her twenty-third year. After our marriage, my wife would sometimes say to me: "You saw my picture first; that is not fair." In truth, I had, and not long after I came to know about the proposal. I cannot deny that it sent my mind into turmoil.

Soon the official *miai*[4] photograph arrived. My mother liked

---

[3]Area south of Kyoto around Wakayama
[4]*Miai* is the name for the first meeting of the parties of the proposed marriage

it and said, "She looks intelligent." Next, I had to have my own picture taken. My mother looked at my closely cropped head, saying, "You have to grow your hair." But hair does not grow that fast, and I had to take the picture with my hair half-grown. It stood up straight, and I could not comb it down. I looked very uncomfortable in the suit that I was unaccustomed to wearing. The photograph was terrible. My wife still teases me about it, "You looked so depressed and sickly in that picture!" It was not only the photograph — her overall impression of me in person was also not very satisfying. I do not care to characterize any further the kind of impression she formed, but I must relate how our first meeting went.

The *miai* was held at the Osaka Hotel, a small, comfortable building at the foot of the Korai Bridge. The following account is from my wife, Sumi Yukawa:

> That morning, I was thinking that I would not wear a flower in my hair, although it was customary in that period for girls to tie back their hair, and to decorate it with an artificial flower. My sister, Nobuko, came and said, "Why don't you wear your flower? You must wear a big beautiful one, especially today; that rose would look best."
>
> I said, "I would not want to marry a man who decides on a marriage because the girl is pretty. If that is his reason, then he will abandon me if I become sick and haggard."
>
> My sister laughed loudly, "How like you! Don't worry about it, just wear the rose!"
>
> I put the rose into my hair. I wore a kimono with spindle designs over a base of black and brown. We left our house at Uchiawaji — my father Genyo, my mother Michi, my sister Nobuko, and I. The Osaka Hotel is beyond the third bridge over the Yokabori

River. My brother, Seiyo, who had already taken over the direction of the hospital from my father, was to go directly from there to the hotel.

Only a few minutes after our arrival, Professor Ogawa entered the room saying, "We are a little late." He seemed cheerful, and we began to relax. Mrs. Ogawa entered just after him, wearing her hair in a foreign style, with bangs and a large chignon. She walked modestly, her large white face tilted downward. Hideki then entered between Kishi-san and his wife, but I was too nervous to observe him closely.

We decided to dine immediately, in another room. My brother, Seiyo, sat across from me and spoke with Hideki, who sat beside him. "With whom do you work at the University?" he asked.

"Professor Tamaki."

He seemed to be trying to say as little as possible, and he spoke in a very faint voice. My brother tried to continue the conversation by finding another topic. "Do you know so-and-so?"

"No."

My brother was perplexed; on the other hand, I gradually became calm and began to observe Hideki. He was wearing a conservative brown suit and a tie with an intricate design. As I had noticed in the photograph, his hair was only partly grown out. Dry hairs were standing in line from the back to the front of his head. His forehead was large, and he wore large black tortoise shell rimmed spectacles over a long pale face.

My wife ends her impression with the following sentences: "I worry a little that he is too quiet, but whatever the appearance, there is no mistaking that he has a studious and brilliant mind. I think that I can trust him with my life." As for my own impressions, she seemed to me to be an innocent and over-protected girl. On the contrary, she was actually much more practical than I. Compared to her, I was a mere puppet of fate; or perhaps, more like a silkworm, wrapped in its chrysalis.

In any case, the business went well. After about a month, my mother and I visited the Yukawa household in Osaka; then they visited us in Kyoto. By that time, the marriage was almost arranged. Sumi, her mother, and I went by automobile to look at an apartment house named "Shigaraki." It is a little unusual for a couple to go to look for an apartment before the wedding, but we had a reason. It was decided that I would be adopted by the Yukawa family upon marriage. It was not an unfamiliar practice to me; both my father and my mother's father had been adopted in that way. Genyo Yukawa also was adopted; his name had been Josaburo Sakabé before that.

Sakabé was the name of a very prominent samurai family. Josaburo's father must have been very obstinate; he remonstrated with his lord, and was sentenced to commit harakiri. The family was abolished, and Josaburo underwent great hardship while he was being raised by his brother's wife. He graduated from Wakayama Kenritsu Normal School and found employment as a teacher at an elementary school in the village of Hiisaki in Hidaka. He impressed the mayor of Hiisaki, Genseki Yukawa, and was adopted as a son. As the Yukawa family traditionally became doctors, Josaburo studied for that profession; he changed his name to Genyo.

The village of Hiisaki is not far from Doseiji Temple, which is famous for the legend of Anchin and Kiyohimé. It is a seaside village northeast of the mouth of the Hidaka River; the Yukawa house faces the sea and has a fine view. Sometimes the village is struck by tidal waves. Genseki served the village both as doctor and as mayor. My mother-in-law is named Michi (meaning "road") because she was born when Genseki, her father, opened a new road for the village.

Genyo later graduated from the Kyoto Medical School and opened a practice in Iyo, in Shikoku. Afterwards, he moved back to Kishu and opened a small hospital in the town of Gobō, near the village of Hiisaki. He seemed to have been a studious person; the book that he published at that time, *The New Book on the Treatment of Gastrointestinal Problems,* apparently sold well, making it possible for him to realize his dream of studying in Germany.

After returning from Europe, he opened a practice in Osaka. He built the hospital at Imabachi Sanchome before Sumi Yukawa was born. It was perceptive of him to open a gastrointestinal hospital in Osaka, which is famous for eating well. The hospital was so popular that he had to examine almost a hundred patients, from early morning until three in the afternoon. He also had to make the rounds of the bed patients, and after that, to make house calls. It was very hard work, and his heart became exhausted. By the time of our marriage, he had retired and was resting at home.

To describe Genyo Yukawa, it would be best to borrow a passage from Soseki Natsume's novel, *Kojin:*

> The director of the hospital usually wore a black morning-coat, and was followed by an assistant and a nurse. He was an impressive man with a dark complexion and a straight nose, who showed dignity in his manner and speech. When Mizawa asked:

> "Is it possible for me to travel? Is there danger of developing an ulcer?" Or, "Was it a good policy to be hospitalized like this?" the director replied:

> "Probably," or some such simple answer.

That passage occurs when a character, named Mizawa, is in a hospital in Osaka. According to Toyotaka Komiga's biography of Soseki, the latter was hospitalized for a stomach ulcer in the forty-first year of Meiji. The novel, *Kojin*, was serialized in the *Asahi* newspaper from the first to the second year of Taisho. It is clear that Soseki was recalling Genyo in this passage from *Kojin*.

However, the time I am writing about is twenty-odd years later. It was, perhaps, January of 1932. The day was cold, and the car carrying the three of us ran through Kyoto under a weak winter sun. Very soon, the car turned into a narrow road in east Saubongi. The place was an old house, whose gate was at the end of a narrow path. From the back of the house, one could see the Kamo River. Two connecting rooms in the back had been newly floored and painted. The white paper of the new shoji-screen was brilliant. When the screen was opened, the East Mountain could be seen straight ahead. The families had decided we should live there after the wedding and Sumi's mother was negotiating with the owner of "Shigaraki."

We two who were not busy, stood at the window, admiring the scenery.

"What a beautiful sight! What do they call that bridge?"

"Kojin Bridge. I cross it every day when I go to the University."

I was a little more talkative than usual, and pointed beyond the bridge: "That's the university clock tower. The physics classrooms used to be close by it."

We liked "Shigaraki," but decided finally not to live there. After the wedding in April, we lived in the house in Uchiawaji, and I went to the University by the interurban train that connects Osaka and Kyoto. Many aspects of my behavior during that period were not normal. My feelings were a little unbalanced, as my life underwent great changes, and not only in my private life. In March, before my wedding, I was told by Professor Tamaki that I was to begin lecturing on quantum mechanics, as an instructor in the Physics Department. Unable to adjust easily to new situations, I was rather upset.

From right: Hideki, his wife, Sumi, mother-in-law, Michi and father-in-law, Genyo Yukawa at their home in Osaka. 1932.

much energy available.

I became increasingly confident. I spoke to everyone about the new theory during the meeting of the Kikuchi research group. Kikuchi said, "If there is such a charged particle, it should become visible in the Wilson cloud chamber, should it not?" I answered, "Yes, the particle can be found in the cosmic rays."

Not long afterwards, in November, I presented the new theory to the Osaka branch of the Physico-Mathematical Society of Japan. Professor Nishina was very interested in the theory, and he congratulated me. I finished a paper in English by the end of November and sent it to the society for publication. The reason that I finished the paper so quickly was that my wife kept urging me, "Please write the English paper and show it to the world."

I felt like a traveler who rests himself at a small tea shop at the top of a mountain slope. At that time I was not thinking about whether there were any more mountains ahead.

Hideki Yukawa and Shoichi Sakata with members of Kikuchi laboratory, Osaka University. Professor Seishi Kikuchi is at the center in the front row. Yukawa is at the left-most and Sakata is at the right-most in the back row. On the roof of the Department of Physics, around 1934-1935.

$8 \times 23 = 184$

$12 \times 31 = 372$

DEPARTMENT OF PHYSICS
OSAKA IMPERIAL UNIVERSITY.

1

DATE Nov. 1, 1934

NO. 1

On the Interaction of
Elementary Particles. I,
By Hideki Yukawa

§1. Introduction

At the present stage of the quantum theory little is known about the nature of interaction between elementary particles. For example, the force acting between a neutron and a proton whether is an ordinary attraction force or an "exchange interaction" "Platzwechsel" interaction first proposed by Heisenberg. Recently Fermi[1] has treated the problem of β-ray disintegration on the hypothesis of "neutrino". According to this theory a neutron and a proton can interact by emitting and absorbing a neutrino and an electron. Unfortunately the energy of interaction calculated on such assumption[2] is much too small to account for the binding of neutrons and protons in the nucleus. To remove this defect we have to modify the theory of Heisenberg or Fermi in the following way.

The transition of a heavy particle from a neutron state to a proton state is not always accompanyed

[1] E. Fermi, Zeits. f. Phys. **88**, 161, (1934).

[2] Ig. Tamm, Nature, **133**, 981 (1934); D. Iwanenko, ibid., 981 (1934).

The front sheet of Yukawa's hand-written draft of the first paper on meson theory with the date November 1, 1934.

On the Interaction of Elementary Particles. I.

By Hideki Yukawa.

(Read Nov. 17, 1934)

§1. Introduction

At the present stage of the quantum theory little is known about the nature of interaction of elementary particles. Heisenberg considered the interaction of " Platzwechsel " between the neutron and the proton to be of importance to the nuclear structure.[1]

Recently Fermi treated the problem of β-disintegration on the hypothesis of " neutrino "[2]. According to this theory, the neutron and the proton can interact by emitting and absorbing a pair of neutrino and electron. Unfortunately the interaction energy calculated on such assumption is much too small to account for the binding energies of neutrons and protons in the nucleus.[3]

To remove this defect, it seems natural to modify the theory of Heisenberg and Fermi in the following way. The transition of a heavy particle from neutron state to proton state is not always accompanied by the emission of light particles, i. e., a neutrino and an electron, but the energy liberated by the transition is taken up sometimes by another heavy particle, which in turn will be transformed from proton state into neutron state. If the probability of occurrence of the latter process is much larger than that of the former, the interaction between the neutron and the proton will be much larger than in the case of Fermi, whereas the probability of emission of light particles is not affected essentially.

Now such interaction between the elementary particles can be described by means of a field of force, just as the interaction between the charged particles is described by the electromagnetic field. The above considerations show that the interaction of heavy particles with this field is much larger than that of light particles with it.

(1) W. Heisenberg. Zeit f. Phys. **77**, 1 (1932); **78**, 156 (1932); **80**, 587 (1933). We shall denote the first of them by I.

(2) E. Fermi, ibid. **88**, 161 (1934)

(3) Ig. Tamm, Nature **133**, 981 (1934); D. Iwanenko, ibid. 981 (1934).

The first page of the paper on meson theory published in *Proceedings of the Physico-Mathematical Society of Japan*, **17**, 27 (1935).

Epilogue

I end my memoir here, at least for now. It has been a record of many things that happened to me, and my reactions to them, from the time I was born until I was twenty-seven years and a few months old. There may be some errors in my memory on some minor points. In order to minimize them, I have tried to listen to as many accounts as possible from friends and relatives. I do not want to write beyond this point, because those days when I studied relentlessly are nostalgic to me, and on the other hand, I am sad when I think how I have become increasingly preoccupied with matters other than study. But there are some acknowledgements that I must make. I wish to express my gratitude to the people who made it possible for me to study physics to the best of my ability. All of the people who figure in this memoir, however different they may be from each other, are such people. There are also many, who are not mentioned here for some reason or other, who must nevertheless be thanked.

I must also write about the people who have helped me in my research and contributed to its development after the period covered in this memoir. I count myself especially lucky to have had outstanding assistants during the infancy of the meson theory, from the ninth to around the fourteenth year of Showa

(1934-1939). Shoichi Sakata came to Osaka University from the Institute for Physical and Chemical Research in Tokyo, and we came to share our research efforts. Two years later, Mitsuo Taketani began to participate in our discussions. Next, Minoru Kobayashi joined our little groups. I am a solitary man, and also obstinate, so I find it amazing that we four could have worked together so harmoniously for so long.

Lastly, I want to mention that this memoir (in its original form) found unexpectedly numerous and varied readers. I received many letters of praise, as well as letters pointing our errors, from people known and unknown to me. Because I could not reply each of these letters, I would like to take this opportunity to thank them all.

On the Interaction of Elementary Particles. I.

By Hideki YUKAWA.

(Read Nov. 17, 1934)

§ 1. Introduction

At the present stage of the quantum theory little is known about the nature of interaction of elementary particles. Heisenberg considered the interaction of " Platzwechsel " between the neutron and the proton to be of importance to the nuclear structure.[1]

Recently Fermi treated the problem of β-disintegration on the hypothesis of " neutrino "[2]. According to this theory, the neutron and the proton can interact by emitting and absorbing a pair of neutrino and electron. Unfortunately the interaction energy calculated on such assumption is much too small to account for the binding energies of neutrons and protons in the nucleus.[3]

To remove this defect, it seems natural to modify the theory of Heisenberg and Fermi in the following way. The transition of a heavy particle from neutron state to proton state is not always accompanied by the emission of light particles, i. e., a neutrino and an electron, but the energy liberated by the transition is taken up sometimes by another heavy particle, which in turn will be transformed from proton state into neutron state. If the probability of occurrence of the latter process is much larger than that of the former, the interaction between the neutron and the proton will be much larger than in the case of Fermi, whereas the probability of emission of light particles is not affected essentially.

Now such interaction between the elementary particles can be described by means of a field of force, just as the interaction between the charged particles is described by the electromagnetic field. The above considerations show that the interaction of heavy particles with this field is much larger than that of light particles with it.

---

(1) W. Heisenberg, Zeit f. Phys. 77, 1 (1932) ; 78, 156 (1932); 80, 587 (1933). We shall denote the first of them by I.

(2) E. Fermi, ibid. 88, 161 (1394).

3) Ig. Tamm, Nature 133, 981 (1934); D. Iwanenko, ibid. 981 (1934).

On the Interaction of Elementary Particles. I.

In the quantum theory this field should be accompanied by a new sort of quantum, just as the electromagnetic field is accompanied by the photon.

In this paper the possible natures of this field and the quantum accompanying it will be discussed briefly and also their bearing on the nuclear structure will be considered.

Besides such an exchange force and the oridinary electric and magnetic forces there may be other forces between the elementary particles, but we disregard the latter for the moment.

Fuller account will be made in the next paper.

§ 2. Field describing the interaction

In analogy with the scalar potential of the electromagnetic field, a function $U(x, y, z, t)$ is introducd to describe the field between the neutron and the proton. This function will satisfy an equation similar to the wave equation for the electromagnetic potential.

Now the eqnation

$$\left\{\Delta - \frac{1}{c^2}\frac{\partial^2}{\partial t^2}\right\} U = 0 \tag{1}$$

has only static solution with central symmetry $\frac{1}{r}$, except the additive and the multiplicative constants. The potential of force between the neutron and the proton should, however, not be of Coulomb type, but decrease more rapidly with distance. It can be expressed, for example, by

$$+ \text{ or } -g^2\frac{e^{-\lambda r}}{r}, \tag{2}$$

where g is a constant with the dimension of electric charge, i. e., $cm.^{\frac{3}{2}}$ sec.$^{-1}$ gr.$^{\frac{1}{2}}$ and λ with the dimention cm.$^{-1}$

Since this function is a static solution with central symmetry of the wave equation

$$\left\{\Delta - \frac{1}{c^2}\frac{\partial^2}{\partial t^2} - \lambda^2\right\} U = 0, \tag{3}$$

let this equation be assumed to be the correct equation for U in vacuum. In the presence of the heavy particles, the U-field interacts with them and causes the transition from neutron state to proton state.

Hideki YUKAWA.

Now, if we introduce the matrices[4]

$$\tau_1=\begin{pmatrix} 0 & 1 \\ 1 & 0 \end{pmatrix}, \quad \tau_2=\begin{pmatrix} 0 & -i \\ i & 0 \end{pmatrix}, \quad \tau_3=\begin{pmatrix} 1 & 0 \\ 0 & -1 \end{pmatrix}$$

and denote the neutron state and the proton state by $\tau_3=1$ and $\tau_3=-1$ respectively, the wave equation is given by

$$\left\{\Delta-\frac{1}{c^2}\frac{\partial^2}{\partial t^2}-\lambda^2\right\}U=-4\pi g\,\tilde{\Psi}\frac{\tau_1-i\tau_2}{2}\Psi, \tag{4}$$

where Ψ denotes the wave function of the heavy particles, being a function of time, position, spin as well as τ_3', which takes the value either 1 or -1.

Next, the conjugate complex function $\tilde{U}(x,y,z,t)$, satisfying the equation

$$\left\{\Delta-\frac{1}{c^2}\frac{\partial^2}{\partial t^2}-\lambda^2\right\}\tilde{U}=-4\pi g\tilde{\Psi}\frac{\tau_1+i\tau_2}{2}\Psi, \tag{5}$$

is introduced, corresponding to the inverse transition from proton to neutron state.

Similar equation will hold for the vector function, which is the analogue of the vector potential of the electromagnetic field. However, we disregard it for the moment, as there's no correct relativistic theory for the heavy particles. Hence simple non-relativistic wave equation neglecting spin will be used for the heavy particle, in the following way

$$\left\{\frac{h^2}{4}\left(\frac{1+\tau_3}{M_N}+\frac{1-\tau_3}{M_P}\right)\Delta+ih\frac{\partial}{\partial t}-\frac{1+\tau_3}{2}M_Nc^2-\frac{1-\tau_3}{2}M_Pc^2\right.$$
$$\left.-g\left(\tilde{U}\frac{\tau_1-i\tau_2}{2}+U\frac{\tau_1+i\tau_2}{2}\right)\right\}\Psi=0, \tag{6}$$

where h is Planck's constant divided by 2π and M_N, M_P are the masses of the neutron and the proton respectively. The reason for taking the negative sign in front of g will be mentioned later.

The equation (6) corresponds to the Hamiltonian

$$H=\left(\frac{1+\tau_3}{4M_N}+\frac{1-\tau_3}{4M_P}\right)p^2+\frac{1+\tau_3}{2}M_Nc^2+\frac{1-\tau_3}{2}M_Pc^2$$
$$+g\left(\tilde{U}\frac{\tau_1-i\tau_2}{2}+U\frac{\tau_1+i\tau_2}{2}\right) \tag{7}$$

---

(4) Heisenberg, loc, cit. I.

On the Interaction of Elemetary Particles. I.

where p is the momentum of the particle. If we put $M_N c^2 - M_P c^2 = D$ and $M_N + M_P = 2M$, the equation (7) becomes approximately

$$H = \frac{p^2}{2M} + \frac{g}{2} \{ \tilde{U}(\tau_1 - i\tau_2) + U(\tau_1 + i\tau_2) \} + \frac{D}{2} \tau_3, \tag{8}$$

where the constant term Mc^2 is omitted.

Now consider two heavy particles at points (x_1, y_1, z_1) and (x_2, y_2, z_2) respectively and assume their relative velocity to be small. The fields at (x_1, y_1, z_1) due to the particle at $(x_2 y_2, z_2)$ are, from (4) and (5),

and

$$\left. \begin{array}{l} U(x_1, y_1. z_1) = g \dfrac{e^{-\lambda r_{12}}}{r_{12}} \dfrac{(\tau_1^{(2)} - i\tau_2^{(2)})}{2} \\[3mm] \tilde{U}(x, y_1, z_1) = g \dfrac{e^{-\lambda r_{12}}}{r_{12}} \dfrac{(\tau_1^{(2)} + i\tau_2^{(2)})}{2}, \end{array} \right\} \tag{9}$$

where $(\tau_1^{(1)}, \tau_2^{(1)}, \tau_3^{(1)})$ and $(\tau_1^{(2)}, \tau_2^{(2)}, \tau_3^{(2)})$ are the matrices relating to the first and the second particles respectively, and r_{12} is the distance between them.

Hence the Hamiltonian for the system is given, in the absence of the external fields, by

$$\begin{aligned} H &= \frac{p_1^2}{2M} + \frac{p_2^2}{2M} + \frac{g^2}{4} \{ (\tau_1^{(1)} - i\tau_2^{(1)})(\tau_1^{(2)} + i\tau_2^{(2)}) \\ &\quad + (\tau_1^{(1)} + i\tau_2^{(1)})(\tau_1^{(2)} - i\tau_2^{(2)}) \} \frac{e^{-\lambda r_{12}}}{r_{12}} + (\tau_3^{(1)} + \tau_3^{(2)})D \\ &= \frac{p_1^2}{2M} + \frac{p_2^2}{2M} + \frac{g^2}{2}(\tau_1^{(1)}\tau_1^{(2)} + \tau_2^{(1)}\tau_2^{(2)}) \frac{e^{-\lambda r_{12}}}{r_{12}} + (\tau_3^{(1)} + \tau_3^{(2)})D, \tag{10} \end{aligned}$$

where p_1, p_2 are the momenta of the particles.

This Hamiltonian is equivalent to Heisenberg's Hamiltonian (1),[5] if we take for " Platzwechselintegral "

$$J(r) = -g^2 \frac{e^{-\lambda r}}{r}, \tag{11}$$

except that the interaction between the neutrons and the electrostatic repulsion between the protons are not taken into account. Heisenberg took the positive sign for $J(r)$, so that the spin of the lowest energy state of H^2 was 0, whereas in our case, owing to the negative sign in front of g^2, the lowest energy state has the spin 1, which is required

---

(5) Heisenberg, I.

from the experiment.

Two constants g and λ appearing in the above equations should be determined by comparison with experiment. For example, using the Hamiltonian (10) for heavy particles, we can calculate the mass defect of H^2 and the probability of scattering of a neutron by a proton provided that the relative velocity is small compared with the light velocity.[6]

Rough estimation shows that the calculated values agree with the experimental results, if we take for λ the value between 10^{12}cm^{-1}. and 10^{13}cm^{-1}. and for g a few times of the elementary charge e, although no direct relation between g and e was suggested in the above considerations.

§3. Nature of the quanta accompanying the field

The U-field above considered should be quantized according to the general method of the quantum theory. Since the neutron and the proton both obey Fermi's statistics, the quanta accompanying the U-field should obey Bose's statistics and the quantization can be carried out on the line similar to that of the electromagnetic field.

The law of conservation of the electric charge demands that the quantum should have the charge either $+e$ or $-e$. The field quantity U corresponds to the operator which increases the number of negatively charged quanta and decreases the number of positively charged quanta by one respectively. $\tilde{U}$, which is the complex conjugate of U, corresponds to the inverse operator.

Next, denoting

$$p_x = -ih\frac{\partial}{\partial x}, \quad \text{etc.,} \quad W = ih\frac{\partial}{\partial t},$$

$$m_U c = \lambda h,$$

the wave equation for U in free space can be written in the form

$$\left\{ p_x^2 + p_y^2 + p_z^2 - \frac{W^2}{c^2} + m_U c^2 \right\} U = 0, \tag{12}$$

so that the quantum accompanying the field has the proper mass $m_U = \dfrac{\lambda h}{c}$.

---

(6) These calculations were made previously, according to the theory of Heisenberg, by Mr. Tomonaga, to whom the writer owes much. A little modification is necessary in our case. Detailed accounts will be made in the next paper.

On the Interaction of Elementary Particles. I.

Assuming $\lambda = 5 \times 10^{12} \text{cm}^{-1}$., we obtain for m_U a value 2×10^2 times as large as the electron mass. As such a quantum with large mass and positive or negative charge has never been found by the experiment, the above theory seems to be on a wrong line. We can show, however, that, in the ordinary nuclear transformation, such a qnantum can not be emitted into outer space.

Let us consider, for example, the transition from a neutron state of energy W_N to a proton state of energy W_P, both of which include the proper energies. These states can be expressed by the wave functions

$$\Psi_N(x, y, z, t, 1) = u(x, y, z)e^{-iW_N t/h}, \quad \Psi_N(x, y, z, t, -1) = 0$$

and

$$\Psi_P(x, y, z, t, 1) = 0, \quad \Psi_P(x, y, z, t, -1) = v(x, y, z)e^{-iW_P t/h},$$

so that, on the right hand side of the equation (4), the term

$$-4\pi g \tilde{v} u e^{-it(W_N - W_P)/h}$$

appears.

Putting $U = U'(x, y, z)e^{i\omega t}$, we have from (4)

$$\left\{ \Delta - \left(\lambda^2 - \frac{\omega^2}{c^2} \right) \right\} U' = -4\pi g \tilde{v} u, \qquad (13)$$

where $\omega = \dfrac{W_N - W_P}{h}$. Integrating this, we obtain a solution

$$U'(r) = g \int\!\!\int\!\!\int \frac{e^{-\mu|r-r'|}}{|r-r'|} \tilde{v}(r')u(r')dv', \qquad (14)$$

where $\mu = \sqrt{\lambda^2 - \dfrac{\omega^2}{c^2}}$.

If $\lambda > \dfrac{|\omega|}{c}$ or $m_U c^2 > |W_N - W_P|$, μ is real and the function $J(r)$ of Heisenberg has the form $-g^2 \dfrac{e^{-\mu r}}{r}$, in which μ, however, depends on $|W_N - W_P|$, becoming smaller and smaller as the latter approaches $m_U c^2$. This means that the range of interaction between a neutron and a proton increases as $|W_N - W_P|$ increases.

Now the scattering (elastic or inelastic) of a neutron by a nucleus can be considered as the result of the following double process : the neutron falls into a proton level in the nucleus and a proton in the latter jumps to a neutron state of positive kinetic energy, the total energy being conserved throughout the process. The above argument, then, shows that the probability of scattering may in some case increase

Hideki YUKAWA

with the velocity of the neutron.

According to the experiment of Bonner[7], the collision cross section of the neutron increases, in fact, with the velocity in the case of lead whereas it decreases in the case of carbon and hydrogen, the rate of decrease being slower in the former than in the latter. The origih of this effect is not clear, but the above considerations do not, at least, contradict it. For, if the binding energy of the proton in the nucleus becomes comparable with $m_U c^2$, the range of interaction of the neutron with the former will increase considerably with the velocity of the neutron, so that the cross section will decrease slower in such case than in the case of hydrogen, i. e., free proton. Now the binding energy of the proton in C^{12}, which is estimated from the difference of masses of C^{12} and B^{11}, is

$$12,0036 - 11,0110 = 0,9926.$$

This corresponds to a binding energy 0,0152 in mass unit, being thirty times the electron mass. Thus in the case of carbon we can expect the effect observed by Bonner. The arguments are only tentative, other explanations being, of course, not excluded.

Next if $\lambda < \dfrac{|\hbar|}{c}$ or $m_U c^2 < |W_N - W_P|$, μ becomes pure imaginary and U expresses a spherial undamped wave, implying that a quantum with energy greater than $m_U c^2$ can be emitted in outer space by the transition of the heavy particle from neutron state to proton state, provided that $|W_N - W_P| > m_U c^2$.

The velocity of U-wave is greater but the group velocity is smaller than the light velocity c, as in the case of the electron wave.

The reason why such massive quanta, if they ever exist, are not yet discovered may be ascribed to the fact that the mass m_U is so large that condition $|W_N - W_P| > m_U c^2$ is not fulfilled in ordinary nuclear transformation.

§ 4. Theory of β-disintegration

Hitherto we have considered only the interaction of U-quanta with heavy particles. Now, according to our theory, the quantum emitted when a heavy particle jumps from a neutron state to a proton state, can be absorbed by a light particle which will then in consequence of energy absorption rise from a neutrino state of negative energy to an

---

(7) T. W. Bonner, Phys. Rev. **45**, 606 (1934).

On the Interaction of Elementary Particles. I.

electron state of positive energy. Thus an anti-neutrino and an electron are emitted simultaneously from the nucleus. Such intervention of a massive quantum does not alter essentially the probability of β-disintegration, which has been calculated on the hypothesis of direct coupling of a heavy particle and a light particle, just as, in the theory of internal conversion of γ-ray, the intervation of the proton does not affect the final result.[3] Our theory, therefore, does not differ essentially from Fermi's thory.

Fermi considered that an electron and a neutrino are emitted simultaneously from the radioactive nucleus, but this is formally equivalent to the assumption that a light particle jumps from a neutrino state of negative energy to an electron state of positive energy.

For, if the eigenfunctions of the electron and the neutrino be ψ_k, φ_k respectively, where $k=1, 2, 3, 4$, a term of the form

$$-4\pi g' \sum_{k=1}^{4} \hat{\psi}_k \varphi_k \tag{15}$$

should be added to the right hand side of the equation (5) for $\widetilde{U}$, where g' is a new constant with the same dimension as g.

Now the eigenfunctions of the neutrino state with energy and momentum just opposite to those of the state φ_k is given by $\varphi_k' = -\delta_{kl}\tilde{\varphi}_l$ and conversely $\varphi_k = \delta_{kl}\tilde{\varphi}_l'$, where

$$\delta = \begin{pmatrix} 0 & -1 & 0 & 0 \\ 1 & 0 & 0 & 0 \\ 0 & 0 & 0 & 1 \\ 0 & 0 & -1 & 0 \end{pmatrix},$$

so that (15) becomes

$$-4\pi g' \sum_{k,l=1}^{4} \tilde{\psi}_k \delta_{kl} \tilde{\varphi}_l'. \tag{16}$$

From equations (13) and (15), we obtain for the matrix element of the interaction energy of the heavy particle and the light particle an expression

$$\cdot gg' \int \cdots \int \tilde{v}(\boldsymbol{r}_1) u(\boldsymbol{r}_1) \sum_{k=1}^{4} \tilde{\psi}_k(\boldsymbol{r}_2) \varphi_k(\boldsymbol{r}_2) \frac{e^{-\lambda r_{12}}}{r_{12}} dv_1 dv_2, \tag{17}$$

corresponding to the following double process : a heavy particle falls

---

(8) H. A. Taylor and N. F Mott, Proc. Roy. Soc. A, **138**, 665 (1932).

Hideki YUKAWA

from the neutron state with the eigenfunction $u(r)$ into the proton state with the eigenfunction $v(r)$ and simultaneously a light particle jumps from the neutrino state $\varphi_k(r)$ of negative energy to the electron state $\psi_k(r)$ of positive energy. In (17) λ is taken instead of μ, since the difference of energies of the neutron state and the proton state, which is equal to the sum of the upper limit of the energy spectrum of β-rays and the proper energies of the electron and the neutrino, is always small compared with $m_v c^2$.

As λ is much larger than the wave numbers of the electron state and the neutrino state, the function $\dfrac{e^{-\lambda r_{12}}}{r_{12}}$ can be regarded approximately as a δ-function multiplied by $\dfrac{4\pi}{\lambda^2}$ for the integrations with respect to x_2, y_2, z_2. The factor $\dfrac{4\pi}{\lambda^2}$ comes from

$$\iiint \frac{e^{-\lambda r_{12}}}{r_{12}} dv_2 = \frac{4\pi}{\lambda^2}.$$

Hence (17) becomes

$$\frac{4\pi g g'}{\lambda^2} \iiint \tilde{v}(r) u(r) \sum_k \tilde{\psi}_k(r) \varphi_k(r) dv \qquad (18)$$

or by (16)

$$\frac{4\pi g g'}{\lambda^2} \iiint \tilde{v}(r) u(r) \sum_{k,l} \tilde{\psi}(r) \delta_{kl'} \tilde{\varphi}_{l'}(r) dv, \qquad (19)$$

which is the same as the expression (21) of Fermi, corresponding to the emission of a neutrino and an electron of positive energy states $\varphi_k'(r)$ and $\psi_k(r)$, except that the factor $\dfrac{4\pi g g'}{\lambda^2}$ is substituted for Fermi's g.

Thus the result is the same as that of Fermi's theory, in this approximation, if we take

$$\frac{4\pi g g'}{\lambda^2} = 4 \times 10^{-50} \text{cm}^3. \text{ erg,}$$

from which the constant g' can be determined. Taking, for example, $\lambda = 5 \times 10^{12}$ and $g = 2 \times 10^{-9}$, we obtain $g' \cong 4 \times 10^{-17}$, which is about 10^{-8} times as small as g.

This means that the interaction between the neutrino and the electron is much smaller than that between the neutron and the proton so that the neutrino will be far more penetrating than the neutron and consequently more difficult to observe. The difference of g and g' may be due to the difference of masses of heavy and light particles.

On the Interaction of Elementary Particles. I.

§ 5. Summary

The interaction of elementary particles are described by considering a hypothetical quantum which has the elementary charge and the proper mass and which obeys Bose's statistics. The interaction of such a quantum with the heavy particle should be far greater than that with the light particle in order to account for the large interaction of the neutron and the proton as well as the small probability of β-disintegration.

Such quanta, if they ever exist and approach the matter close enough to be absorbed, will deliver their charge and energy to the latter. If, then, the quanta with negative charge come out in excess, the matter will be charged to a negative potential.

These arguments, of course, of merely speculative character, agree with the view that the high speed positive particles in the cosmic rays are generated by the electrostatic field of the earth, which is charged to a negative potential.[9]

The massive quanta may also have some bearing on the shower produced by cosmic rays.

In conclusion the writer wishes to express his cordial thanks to Dr. Y. Nishina and Prof. S. Kikuchi for the encouragement throughout the course of the work.

Department of Physics,
Osaka Imperial University.

(Received Nov. 30, 1934)

Erratum

On page 55 line 6 "The intervation of the proton" should read "The intervention of the photon".

---

(9) G. H. Huxley, Nature **134**, 418, 571 (1934); Johnson, Phys. Rev. **45**, 569 (1934).

Chapter 12

Kurakuen

I had lived in Kyoto ever since I was aware of my surroundings, but I knew next to nothing about the city of Osaka. I became a resident of Osaka because of my marriage. The train station at Umeda was small and cluttered, but it had an attraction quite different from the station at Kyoto. In it I felt the liveliness that fills all of Osaka. Large old houses lined the streets of Uchiawaji. The Yokobori River flowed steadily to the west, people continuously crossing its numerous bridges. Everyone moved briskly; in contrast to Kyoto's natural beauty, there were lively human beings here. One reason why I decided to live in Osaka was that I wanted to let myself be changed by this new environment.

Earlier that year, my companion in the journey toward truth, Sin-itiro Tomonaga, had left for Tokyo. He was to work in the newly established Nishina Laboratory at the Institute for Physical and Chemical Research (*Riken*). Even I, not a social person and admitting to being solitary, felt rather forlorn about our separation. However, it became another reason to change to a new environment.

I was pursued by many things that I was obliged to do in March and April, and felt like a traveler who must hurry along, carrying a large suitcase hastily crammed with possessions. The wedding took place on the third of April, and there was no time for a honeymoon, or peace of mind enough to plan one. I was to begin lecturing as soon as the new academic quarter began. We had only enough time to go on a one-day trip from Osaka to Wakayama — almost an ordinary outing.

My father-in-law, Genyo, had become even weaker, perhaps because of the strain of the wedding, and after the ceremony he went to Shinwakaura to rest. Two or three days later, we traveled to Wakayama to visit him. He had a back room in the Bokairo Hotel in Shinwakaura. Through his window one could see the sea and had a view of a large rock named Horaiiwa. He was pleased with the French-style soup that Sumi prepared for him, and he also felt happy about our marriage.

I learned that the cherry blossoms were in full bloom, back in the mountains. It began to rain, and we borrowed an umbrella from the hotel in order to walk about in the neighborhood. We returned to dine with my father-in-law as it began to pour. However, I was eager to see the temple of Kimii, which is famous for its cherry trees. It was located at the top of a long flight of stone steps, which I climbed quickly, as I was used to doing. Then I turned and saw my wife struggling up the steps in her violet coat, wearing geta on her feet; she was having a hard time catching up with me. I was no longer a lonely traveler; now I had a companion to take care of — and one who would take care of me. The cherry blossoms were in full bloom.

The spring vacation passed quickly; an announcement reading: "Quantum Mechanics, Lecturer Yukawa, opens April ..." was posted outside the office of the Physics Department. The students did not know that my name had been changed during the spring break and were asking, "Who is this Yukawa? I have never heard of him."

Among the students who heard me give my first lectures were Shoichi Sakata and Minoru Kobayashi, who were in their second year. They were the most enthusiastic students and acquired the deepest understanding of quantum physics. Among those in the next class was Mitsuo Taketani, whom I noticed early in the year. These three students were to become the most invaluable collaborators in my subsequent research.

However, not even those interested persons were impressed by my lecturing. For instance, according to Taketani:

> There were no particular characteristics to Yukawa-san's lectures, which followed Dirac's textbook for the most part. His voice was as gentle as a lullaby and he spoke with little emphasis — it was ideal as an invitation to sleep.

Kobayashi added that my voice was very soft, and that when I spoke, I addressed the blackboard, which made my words hard to understand. In later years, when I gave lectures in foreign lands, I was usually asked to speak more loudly.

As far as I was concerned, something more preoccupying than my lectures were happening. The introduction of quantum mechanics six years earlier had caused a turmoil in physics, which was essentially over. For some time, things had been relatively peaceful, but suddenly another period of turmoil began, one that I was involved in. The year 1932 was even more turbulent for physics than it was for my personal life. Events, each of which, taken alone, could be called revolutionary, occurred three in a row: first, there was the discovery of the neutron; second, the discovery of the positron; third, the atomic nucleus was broken up by artificial means, namely by the use of the particle accelerators. The discipline that is now called nuclear physics was until then a minor branch of study, but because of those three events, it suddenly became the mainstream.

Each of those three developments was a major event in its own right, but the one that had the greatest importance for theoretical physics was the neutron discovery. Theoretical physicists, who had been trying without success to construct a model of the nucleus using only electrons and protons, immediately became animated. The third particle (or the fourth, if the photon is included) called the neutron, was a key to solving the mystery of the atomic nucleus. The idea that the nucleus can be visualized as a collection of neutrons and protons must have occurred to a large number of physicists at the same time, but it was Heisenberg who systematically developed the new nuclear structure. I realized the importance of his work and wrote an introduction to his papers in the *Proceedings of the Physico-Mathematical Society of Japan*. At the same time, I decided to carry the theory one step further.

The problem that I focused on was that the nature of the forces that act upon the neutrons and protons making up the nucleus — that is, the nature of the nuclear forces. By confronting this difficult problem, I committed myself to long days of suffering. In fact, the period from the autumn of the seventh year of Showa to the autumn of the ninth year (1932 to 1934) were the most difficult years of my life. The fact that I was suffering, however, was very satisfying to me; I felt like a traveler carrying a heavy burden and struggling up a slope. I experienced the sorrows and joys of a scientist, and I also had the valuable experiences of a person, living in a home and in a society.

There were many differences between the Kyoto that I was used to and the city of Osaka. One important contrast was that the air of Osaka was much drier; that might be why I began to eat more. My wife, who considered me pallid, was relieved to see my color improve. However, the air of Osaka could not be called good. Cinders from the numerous factories entered the house at Uchiawaji; the veranda became rough with ashes if the glass doors were left open, even slightly. My mother-in-law was fond of cleanliness and she hated to have her *tabi* socks turn black at the back. The maids were kept busy cleaning the veranda. The trees in the garden looked unhealthy and were dark in color, not like the lovely green of the foliage of Kyoto.

Inside, however, the house at Uchiawaji was always very neatly kept, unlike our house at Tōnodan where books overflowed every room. The house at Uchiawaji seemed strangely quiet to me, who was used to my brothers' and my father's loud voice. My father-in-law, on the other hand, would become uncomfortable after speaking for only a little while. He usually sat in one place all day; sometimes he coughed painfully, walking was burdensome to him. He had damaged his heart by working too hard during the last years.

My father-in-law had once been plagued by the vocalizing of a soprano, who lived next door to him during his stay in Germany. As a result, his tastes turned entirely Japanese, or more precisely, Oriental, upon his return. He collected *suzuri* (inkwells) and Oriental paintings; he became interested in the tea ceremony and he learned *nanga* and also *gidayu* singing. He made the members of his family study *nanga* and other Japanese arts. My wife, Sumi, had studied Japanese court dancing of the Yamamura school since she was four years old.

The atmosphere of the house was radically different from that of Tōnodan. To me, laboring in the most modern field of science, the atmosphere was relaxing, the complete opposite of my work: it gave relief to my tired and strained mind. Moreover, my way of thinking changed in that new environment. I gradually became liberated from the stubbornness that would allow only a single absolute. At the same time it awakened initiative and activity that had previously been lacking in me.

Thinking back on this, I realize that my mind must have been slightly unbalanced. It had been three years since I graduated from the University, and what had I done? I had increased my knowledge, but I had not done anything creative. Had I performed as a physicist should? I was a little desperate, although my wife and her parents were completely satisfied that

I was working hard. In retrospect, I was extremely lucky on one occasion, when my father came to visit us at Uchiawaji.

He suggested to my father-in-law, "Why not send Hideki to a foreign country to study?" The reply was, "I will think about it." At that period there was no disadvantage in using Japanese currency, and indeed, after the First World War, Japanese currency had a relatively high value because of the great inflation in Europe, especially in Germany. Therefore, to study in foreign countries using private financial resources was not rare in those days. I rejected that proposal, because I did not want to go to foreign lands until I had finished a work that I could call my own. I wanted to find my own topic of research and to pursue it as far as possible. I did not care how many times I failed. Only if I succeeded would I go to talk with foreign scientists.

My environment changed, and there was also beginning to be a change in my heart. The window that I had shut against society was gradually opening, although I still had trouble expressing myself adequately; I never spoke with my adopted parents unless it was absolutely necessary. I neither disliked nor feared them; it was only that my life-long habit of silence was not easy to break.

It was tiring to have to travel to Kyoto to study, and in addition, my new personal life was trying to my nerves. I developed a relatively mild case of insomnia. If the bedroom door made a slight noise, I had to get up to make sure it was closed. If there was a sound outside the house, I would mumble, "What could that be?"

"It is only a cat."

"No, it must be something heavier."

I was not able to sleep, even after the noise had stopped. Next day, I would decide to change to another bedroom; still, I was unable to sleep. I tried a larger room on the following day, and repeated this until I had tried almost every room in the large house. My adopted parents never said a word about all this — no, my father-in-law finally told my wife, "There are no more rooms."

That is the way my thinking went. It was a kind of selfishness, perhaps it was vanity, perhaps it was arrogance. Yes, it can be called that. What I feared most was being forced to consider a problem that did not interest me, whether in Japan or elsewhere. I wanted to put into research everything I had — knowledge, passion, will. I was a troubled human being who could not work in a half-hearted manner.

The first year of my marriage passed uneventfully, at least externally. During that year, I took the Keihan interurban train each morning from Tenmaubashi Station. Unlike today, the train ran slowly on the winding rails; the view, however, was the same. The thoughts that came to me, as I watched the familiar view, concerned the problem of nuclear forces. How could I find the key to its solution? How was the nuclear force related to the other known forces?

Most forces in nature are not primary. For example, the forces of molecular attraction, the combinatory forces of chemistry, are rather complex in nature. They could not be understood for a long time, but after the appearance of quantum mechanics, it became apparent that molecular attraction and chemical binding are secondary forces, derived from the primary electromagnetic attraction or repulsion acting between the nucleus and the electrons and among the electrons, respectively. Thus, after the introduction of quantum mechanics, the only forces recognized as primary were universal gravitation and electromagnetism.

Both of these fundamental forces are expressed as *fields of force*. The field corresponds to the distribution of the intrinsic characteristics of the force over the points of space. If the field is known at a point, one can predict what force will be felt by something occupying that point. For example, if the electric and magnetic fields are known at a certain point, then one knows the force that would be felt by a particle at that point. Of course, the force depends also on the charge, direction, and speed of the particle.

Then the question arises: Is the new nuclear force a primary one? Or is it a secondary one, derived from the gravitational and electromagnetic forces? That is the fundamental question. Now, gravity cannot be applicable in this case, because the gravitational force between such minuscule objects as protons and neutrons is unimaginably small. It is much too weak to bind together such a compact and strongly aggregated composite as a nucleus. Electromagnetic forces, on the other hand, are much stronger than gravitation; but they are still too weak to be the source of the nuclear force. Not only that: electromagnetism appears to give only forces of repulsion, rather than attraction. Because the neutron is electrically neutral, there should not be any large electrical force between it and other particles; and protons repel each other!

In any case, something very strange must intervene if the nuclear force could be thought of as derived from electro-magnetic forces, as a secondary force. Therefore, it seemed likely that the nuclear force was a third fundamental force, unrelated to gravitation or electromagnetism. Then, perhaps the nuclear force could also find expression as a field.

I had this idea of a nuclear force field very early. Looked at from the quantum mechanical viewpoint, a field of force, almost by necessity, implies that there is a particle accompanying that field. We actually infer the existence of the photon as the particle accompanying the electromagnetic field. Stated this way, the answer appeared almost at hand, but my brain did not work so quickly. I had to take a wrong path first, before I could arrive at my destination.

Those who explore an unknown world are travelers without a map; the map is the result of the exploration. The position of their destination is not known to them, and the direct path that leads to it is not yet made. There may be trails that are left by earlier explorers. If we follow them, will we reach our destination, or must we open an entirely different path? It is after we have arrived that we say, "We really took the hard way to get here!" It is not so hard to find the straight route afterwards, but it is difficult to find one's destination while one is opening the new path, perhaps even starting off in the wrong direction. In retrospect, I was very close to my destination in 1932. Had I pursued the notion of the field of nuclear force, and applied quantum mechanical reasoning, I should have been able to come up with the idea of the meson.[1] Instead, I spent the next two years searching in the dark.

What important information did I possess at the start? In the first place, there exist in this universe four fundamental particles called proton, neutron, electron, and photon. Matter, in the ordinary sense, is made of protons, neutrons, electrons; the electromagnetic field is an aggregation of photons. That is to say, electromagnetic interaction consists of an exchange of photons. For example, to say there is an electrical attraction between a proton and an electron is to say that there is a continuous exchange of photons between the two particles. Protons and electrons play "catch" with the photons. Those were the only particles known at that time — no, there was one more. The positron had just been discovered, but it is not entirely different from the electron. It is the "hole in the sea of electrons," predicted by P.A.M. Dirac — in other words, the *antiparticle* of the electron. When one considered the electron, one was led to the necessity of the positron.

Secondly, the atomic nucleus is made of protons and neutrons; the forces that act between these particles are stronger than electromagnetic forces. Then, if one visualizes the nuclear

[1] in the original, "mesotron."

force field as a game of "catch" between protons and neutrons, the crux of the problem would be the nature of the "ball" or particle. At this point, the electron was nominated as the most likely candidate for that honor. It was not only that there was no known candidate other than the electron (in the wider sense that includes the positron). The electron seemed to be a powerful candidate because electrons not only went around the nucleus in atoms, but sometimes they actually jumped out from the nucleus. The true identity of the beta ray from a radioactive nucleus is that of an electron or positron. Heisenberg had already gone on record with his opinion that the "ball" in the game of nuclear "catch" was the electron. At the beginning, I also decided to follow this line of investigation.

In April 1933, the Physico-Mathematical Society of Japan held a meeting at Tohoku University in Sendai. On this occasion I gave my first research report, on the subject "The electrons within nuclei." I did not have very much confidence in this research and did not, in the long run, publish the paper in the journal. There were many obvious difficulties in treating the electron as the "ball" exchanged between neutron and proton. In the first place, the electron's characteristics, such as its spin and the kind of statistics it obeys, make the electron unsuitable for this role. Nevertheless, I tried to use the electron field that satisfies Dirac's wave equation as the field of nuclear force.

Perhaps it was not very surprising that, after my talk, Nishina suggested that I consider using an electron with other statistics, namely the Bose-Einstein statistics. However, I was still strongly committed to the conservative desire to understand nature in terms of the known particles. Furthermore, if there were some electrons that were different from the normal ones, then they should have been found experimentally. Because of these worries, I could not move ahead rapidly.

During the same meeting, I met Professor Hidetsugu Yagi, and that turned out to be a turning point in my life. Professor Yagi had been teaching electrical engineering for some time at

Tohoku University, but had just been invited to head the Physics Department at the newly built Osaka University. His family had already moved to Osaka and he was living alone in a large house that he rented from a friend. My oldest brother was an instructor at Tohoku University, and he had the idea that I should work at Osaka University, because I was living in Osaka. Upon my brother's introduction, I went to visit Professor Yagi.

I was shown into an impressively large living room; it was twilight, and the room was getting dark. I, a nameless student, could not imagine what sort of person Professor Yagi was; I worried about what I should say. There were footsteps in the hallway, and the professor entered. He began to speak immediately about the condition of the Physics Department at Osaka University. As I listened, I developed a deep trust in this man and decided I would join his department at Osaka.

The first president of Osaka University was Hantaro Nagaoka, whom I greatly admired. There were not many professors in the newly established Faculty of Natural Sciences, and there were no buildings. I was to be an instructor beginning in May, and so I was given a desk in one corner of the Shiomi Institute of Chemical Research, next to the university hospital, to the north of the Tamino Bridge. It was close to the Osaka railroad station and very noisy. I could hear the distressed yelping of dogs used for medical research, housed nearby. It was a different environment from that of Kyoto University, where the animal sounds came from grazing mountain sheep.

To go back a little in the story: in April of the eighth year of Showa, shortly after I returned from the meeting at Sendai, my oldest son Harumi was born. I, too, had become a parent. I also entered an entirely different research organization when I went to Osaka University. The newly organized Faculty of Natural Sciences was a kind of melting pot. Although most of its members were from Tokyo University, some came from other universities. Many of the professors were only five or six years older than I was, and that provided a youthful atmosphere that was absent in the older universities.

Tokiharu Okaya and Josaburo Asada came from The Shiomi Institute as professors of physics, joining Professor Yagi. Okaya's specialty was relativity; I was supposed to associate mainly with him. However, because I was not very interested in relativity at that time, I did my work without much contact with Professor Okaya. Soon, Susumu Tomochika was transferred to Osaka as a professor, while Seishi Kikuchi, who was to be responsible for experiments in nuclear physics, visited occasionally.

In the new natural sciences building, under construction south of Tamino Bridge next to the medical school, there was a room with a high ceiling, occupying both the basement and the first floor. In that room, a Cockroft-Walton nuclear accelerator was to be installed, and experiments were to be performed under Kikuchi. I joined the nuclear research group that was being formed, and it was very stimulating to know that experiments would take place simultaneously with the theoretical studies.

In that same year, in summer, our family moved to a new house in Kurakuen. That house became for me the unforgettable house. Not many people know the name Kurakuen, although it was once popular as a resort during the Taisho era. By 1933 it had become deserted, but it retained the nostalgic atmosphere of a place that looks to its past. To get there, one changed trains at Shukugawa and took a branch line to the Kurakuen station. The bus then took fifteen minutes to pass along pine forests and farmers' fields until it reached a slope. In the hills, near the eastern end of Mt. Rokko's ridge, some houses could be seen: that was Kurakuen.

We had lived there in a rented house the previous summer, partly for the sake of my father-in-law's health, and partly because my mother-in-law hated the everpresent dust and smoke at the house in Osaka. It was a fortunate move for my small family as well. The house was on a southern slope and the air was dry; it was cooler than expected. My father-in-law liked living there and decided to build a house on an empty lot near the bus terminal.

The new house had a magnificent view. My father-in-law, who could not move around very much because of his heart condition, sat near the south window all day and looked at the sea, far away. Often, after dinner, we would sit by the window and look at the lights of Nishimia and Amazaki and at the night trains as they passed by.

From Kurakuen I commuted to both Kyoto University and Osaka University. My research continued during this period, but made no noticeable progress. In retrospect, the basic concept of the meson came to me several times, but its appearance were like flashes of lightning that illuminated darkness for only an instant. Whatever was needed to prolong this flash of light was not yet present in my mind.

I took strolls around Kurakuen on Sundays. My wife was usually busy with the baby and seldom went outside. In front of the house there was a row of cherry trees; to the southwest of the house was a pond surrounded by pine trees. An old-fashioned European-style building of red brick was visible — the Kurakuen Hotel. In the past, it attracted people who wished to escape the heat of summer, offering cool baths of spring water. There was a period when famous personages stopped at that hotel, but by the time I came upon it, it was completely ruined. Ivy crept upon the bricks, but there was no other sign of life.

When I climbed the slope to the northeast, the trees became scarce; rocks were bare, and the view widened. At the top of the hill there was a large pond, whose pure blue waters were undisturbed. Its color contrasted beautifully with the white rock facing. There was a stone building on the other side of the pond, a circular European-style building that looked at first like an old castle; its reflected image in the pond was very clear. I felt as though I was in a story out of the Brothers Grimm, like those I loved so much in my early school years. I imagined that a witch might live in that castle, or that a kidnaped princess might be sleeping there. When I approached the circular building, I noticed that the entrance doors were missing and that there was no sign of life. Inside were only the remains of some picnicker's lunch. Apparently the building was started with the intention of making it a hotel, but the project was never completed. There was not even a suggestion of life. I climbed to the second floor and enjoyed the exotic atmosphere for a while.

In May, our house was beautiful with the cherry trees in full bloom. In September, we found mushrooms among the pine trees. I enjoyed my strolls greatly in Kurakuen, but no new ideas came to me.

The ninth year of Showa, (1934) came around. The new natural sciences building was finished and I was to move into that impressive three-story structure in April. The street ran directly in front of the building; it led to the train station, and was always very busy and full of trucks. I felt as though I had to work hard in that building; at times I felt as though I was being pursued. I began to be impatient that my research had no definite results.

I resigned from my position at Kyoto University, becoming a full-time instructor at Osaka University. I was to begin lecturing on electromagnetism in a new academic quarter, although it was not my favorite subject. My mind, instead, remained concentrated on the problem of the nuclear force. One day, among the newly arrived periodicals, I found Fermi's paper concerning beta decay, I think I must have grown pale as I read it. Had I been beaten by Fermi a second time? I thought so, for the reasons that follow.

Within a heavy nucleus, a neutron becomes a proton and an electron is ejected, or conversely, a proton becomes a neutron and a positron is ejected — that had been the picture of beta decay. But there was a huge defect in this concept, for if the electron (or positron) is ejected alone, then the law of conservation of energy is found to be violated in the process. There had been many arguments on that point for several years. Bohr was of the opinion that it was, indeed, true that energy was not conserved. One reason why I could not use the electron as the "ball" for nuclear forces was related to that point.

However, in 1931 Pauli had presented the following idea to a conference: the electron (or positron) perhaps does not emerge unaccompanied from the nucleus during beta decay. Instead, something called the neutrino emerges with it, and that particle carries the energy necessary for the conservation law to be valid. I was not aware of Pauli's argument; Fermi, however, had based his theory of beta decay on Pauli's idea. After reading Fermi, I wondered whether the problem of the strong nuclear force could not be solved in the same manner. That is to say, could the neutrons and protons be playing "catch" with a *pair* of particles, namely an electron and a neutrino? The "ball" would be replaced by a pair of particles.

By the time I started to think seriously along those lines, several foreign scientists were working on the same idea, and their research appeared in the journals. However, the results were negative; the force resulting from the exchange of the pair, electron and neutrino, was incomparably smaller than the nuclear force. I was heartened by the negative result, and it opened my eyes, so that I thought: Let me not look for the particle that belongs to the field of the nuclear force among the known particles, including the new neutrino. If I focus on the characteristics of the nuclear force field, then the characteristics of the particle I seek will become apparent. When I began to think in this manner, I had almost reached my goal.

I was unable to produce creative ideas during the day, getting lost in the various equations written on pieces of paper. On the other hand, when I lay down in bed at night, interesting ideas entered my head. They seemed to grow, unhampered by the rows of equations. Then I became tired, and eventually fell off to sleep.

When I thought about those ideas the next morning, I found that they were all worthless. My hopes disappeared with the morning light, as if by the hand of the devil. I don't know how many times that experience was repeated!

On the twenty-first of September, 1934, a strong wind was blowing in the morning when, as usual, I left the house for the university. The wind was terribly strong, carrying many objects aloft as though they were leaves. By the side of the road were fallen trees. I turned back, thinking it dangerous to continue walking. My wife was pregnant and close to term with our second child. The midwife came to our house as soon as the wind let up a bit. That typhoon, the famous Muroto typhoon, passed and the weather suddenly turned cool; sunny autumn days followed. My second son, Takaki, was born on the twenty-ninth. His brother was then one-and-a-half years old. I was trying to sleep in a small room in the back of the house, but as usual, I was thinking; my insomnia was back again. Beside my bed lay a notebook, so that if I had an idea, I could write it down. That went on for several days.

The crucial point came to me one night in October. The nuclear force is effective at extremely small distances, on the order of 0.02 trillionth of a centimeter. That much I knew already. My new insight was the realization that this distance and the mass of the new particle that I was seeking are inversely related to each other. Why had I not noticed that before? The next morning, I tackled the problem of the mass of the new particle and found it to be about two hundred times that of the electron. It also had to have the charge of plus or minus that of the electron. Such a particle had not, of course, been found, so I asked myself, "Why not?" The answer was simple: an energy of 100 million electron volts would be needed to create such a particle, and there was no accelerator, at that time, with that